Land Use Impacts on Climate

Land Use Impacts on Climate

Editors

Manoj Negi and Jyoti Monga

scitus academics

Land Use Impacts on Climate

Edited by **Manoj Negi and Jyoti Monga**

Printed in 2017

ISBN: 978-1-68117-145-6

Library of Congress Control Number: 2015936562

SCITUS Academics LLC,
616, Corporate Way, Suite 2, 4766,
Valley Cottage, NY 10989

www.scitusacademics.com

Contents

Preface

Since the industrial revolution, climate changes caused by greenhouse gas emissions have attracted the attention globally from a diverse community including both politicians and scientists; however, the pending issues on the research of the climatic impact of land cover and land-use changes via terrestrial bio geophysical processes is far from resolved. The scientific community needs to provide comprehensive scientific support for global change adaptation. Land cover and land use changes directly lead to change in the surface condition, thereby changing the surface albedo and surface roughness, and then affects surface heat balance and water cycles. In addition, the urban expansion causes aggravation of urban heat island effect. It is of significance to understand climatic impacts accompanied by bio geophysical processes induced by land-use changes, which is critical for food safety, extreme climate, biodiversity protection, and a series of problems. Land use change, in turn, is occurring in the context of climate changes at a variety of scales.

Editor

Chapter 1

Additive Threats from Pathogens, Climate and Land-Use Change for Global Amphibian Diversity

Christian Hof[1, 2], Miguel B. Arau´jo[1, 2, 3], Walter Jetz[4], and Carsten Rahbek[1]

[1]Center for Macroecology, Evolution and Climate, Department of Biology, University of Copenhagen, Universitetsparken 15, 2100 Copenhagen Ø, Denmark

[2]Department of Biodiversity and Evolutionary Biology, National Museum of Natural Sciences, CSIC, C/Jose´ Gutie´rrez Abascal 2, 28006 Madrid, Spain

[3]Rui Nabeiro Biodiversity Chair, CIBIO, University of E´ vora, Largo dos Colegiais, 7000 E´vora, Portugal

[4]Department of Ecology and Evolutionary Biology, Yale University, 165 Prospect Street, New Haven, Connecticut 06520-8106, USA

ABSTRACT

Amphibian population declines far exceed those of other vertebrate groups, with 30% of all species listed as threatened by the International Union for Conservation of Nature[1, 2, 3]. The causes of these declines are a matter of continued research, but probably include climate change, land-use change and spread of the pathogenic fungal disease chytridiomycosis[1, 4, 5]. Here we assess the spatial distribution and interactions of these primary threats in relation to the global distribution of amphibian species. We show that the greatest proportions of species negatively affected by climate change are projected to be found in Africa, parts of northern South America and the Andes. Regions with the highest projected impact of land-use and climate change coincide, but there is little spatial overlap with regions highly threatened by the fungal disease. Overall, the areas harbouring the richest amphibian faunas are disproportionately more affected by one or multiple threat factors than areas with low richness. Amphibian declines are likely to accelerate in the twenty-first century, because multiple drivers of extinction could jeopardize their populations more than previous, mono-causal, assessments have suggested.

MAIN

Amphibians are experiencing population declines in all regions of the world[2, 6]. Causes for this global decline have been identified. Among the highest ranking threats are anthropogenic land-use changes, leading to habitat destruction and fragmentation, and the fatal disease chytridiomycosis, which is caused by the chytrid fungus *Batrachochytrium dendrobatidis*. Other threats include climate change, which may interact with chytridiomycosis, environmental pollution, direct exploitation for the food, medicine and pet trades, increase in ultraviolet-B irradiation due to anthropogenic ozone depletion, and the spread of invasive species[4, 5].

Many studies have assessed how these threats affect amphibian populations and how they may interact at local and regional scales[7, 8, 9, 10]. Recent assessments have used bioclimatic envelope models to project climate change impacts on amphibian diversity on a continental scale[11, 12]. Attempts have also been made to assess the relative importance of different threats for large groups of species[13, 14], but not in a spatially explicit way. Several hypotheses have suggested that there may be interactions of chytridiomycosis with land-use change[15] and climate change[9], yet no final consensus has been reached[10, 16]. Preliminary models of the global geography of chytridiomycosis under climate change have been provided[17], but an integrative, analytical, spatially explicit assessment at a global scale of the spatial interactions of the most severe threats is urgently needed.

Using a nearly complete global data set of 5,527 amphibian species, we elucidate how the spatial interaction at the global scale of three important threats (climate change, chytridiomycosis and land-use change, see Methods) could affect global amphibian diversity between a baseline period (1980) and late in the present century (2080). Threat from future climate change was estimated as the proportion of species locally losing climatic suitability ('climate losers') per area, as given by species-specific bioclimatic models (see Methods). Because the spatial distribution of species richness varies considerably among the three amphibian orders Anura (frogs and toads, 'frogs' hereafter), Caudata (salamanders and newts, 'salamanders') and Gymnophiona ('caecilians') (Fig. 1a), we conducted climate change analyses separately for each group. Threat from chytridiomycosis was quantified as the future probability of occurrence of *B. dendrobatidis* from a bioclimatic model projection[17]. Lastly, estimates of future land-use change (that is, changes from natural to human-encroached land cover) were based on the projections of the Millennium Ecosystem Assessment[18, 19, 20].

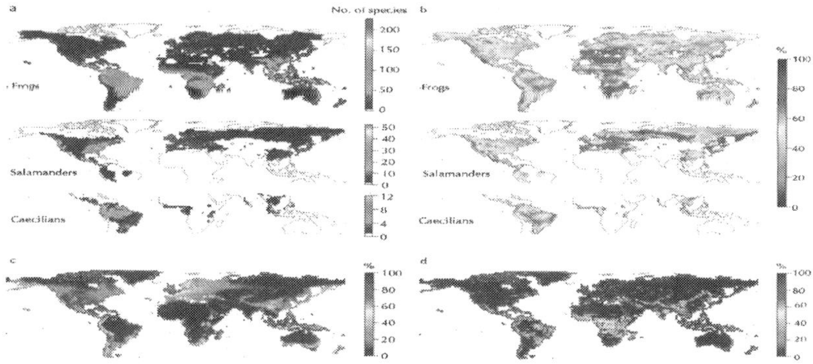

Figure 1: Current amphibian species richness and the intensity of three factors threatening global amphibian diversity projected for the

year 2080. a, Spatial variation of species richness (number of species per grid cell) of frogs (n = 4,875), salamanders (n = 508) and caecilians (n = 144). b, Intensity of threat from climate change, given as the proportion of species projected to lose climatic suitability in a given area (arithmetic mean across 14 GCMs, 3 emission scenarios and 3 modelling algorithms). c, Intensity of threat from chytridiomycosis, given as the projected probability of occurrence of *B. dendrobatidis* (arithmetic mean across 3 GCMs and 2 emission scenarios, data from ref. 17). d, Intensity of threat from land-use change, given as the proportion of a given area projected to be converted from a natural to an anthropogenic state (arithmetic mean across 4 scenarios, data from ref. 19). For further details on the quantification of threat intensities, see Methods. White areas in panels a and b indicate the absence of a given amphibian order.

The three major global threats to amphibians exhibit characteristic, yet disparate geographic patterns (Fig. 1b–d). For frogs, regions with a high projected impact of climate change (regions where high proportions of climate losers in 2080 coincide with a high level of species richness) are the northern Andes and parts of the Amazon and the Cerrado in South America, large areas of sub-Saharan tropical Africa, and a small region in South East Asia (Fig. 1b). In the northern Andes, which harbour the greatest frog diversity worldwide, the proportion of probable climate losers reaches 166 species (73% of the local frog fauna). Globally, the proportion of frogs likely to become climate losers measured as average per grid cell is 54% (±10% (standard deviation, s.d.)). For salamanders, western North America, northern Central America and southern and south-eastern Europe are the regions projected to be most heavily affected by climate change, as are some areas in northern South America for caecilians (Fig. 1b). In several areas of Central America, up to 21 species (66% of the local salamander fauna) are projected to lose climatic suitability (global mean ~56% ± 15% s.d.) (Fig. 1b). The regions with the highest projected probability of occurrence of chytridiomycosis are located in mostly temperate climates as well as mountainous and coastal regions (Fig. 1c). Areas with high projected land-use change are mainly found in tropical Central and South America, tropical Africa and montane parts of central and southern Asia (Fig. 1d).

Geographic coincidence in the intensity of the three types of threat is highly uneven and varies strongly among the three amphibian orders (Figs 1b–d and 2). Within the range occupied by frogs, the spatial overlap between the top 25% affected grid cells for the three threat types is small: 6.1% of these cells (out of a 25% possible) overlap for chytridiomycosis and climate change, 9.1% for climate change and land-use change and 8.6% for chytridiomycosis and land-use change (Figs 2 and 3). The small degree of overlap between areas of highest impact from chytridiomycosis and climate change arises because of the strong association of *B. dendrobatidis* with humid and cool conditions found in temperate regions[21] and cool tropical high mountains, whereas the negative impacts from climate change are more prominent in the warm and wet tropics (see Fig. 1a). For salamanders, the comparable figures for spatial overlap are 12.2%, 4.6% and 10.0%, respectively (Fig. 3). Globally, more than half of the distributional area of each of the three amphibian orders is likely to be highly affected by at least one of the three major threat factors. When only regions with the highest species richness are considered, then about two thirds of the areas with the richest frog and salamander faunas (half of the area for caecilians) are projected to become heavily affected by 2080 (Fig. 4).

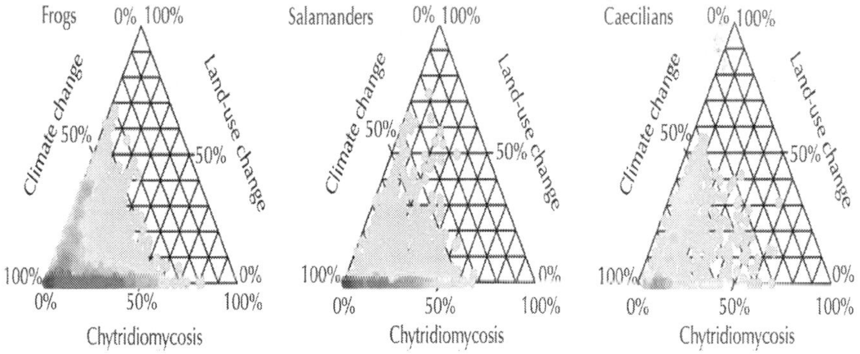

Figure 2: Relationships among the intensities of the three main factors threatening global amphibian diversity. Each point refers to one grid cell in Fig. 1 and represents the relative intensity of each of the three threats when considered jointly: climate change (proportion of species losing climatic suitability); chytridiomycosis (probability of occurrence of chytridiomycosis); and land-use change. For details on the quantification of threat intensities, see Fig. 1 and Methods. Values of threat intensity are standardized to vary between 0

and 100%, and then transformed into relative proportions to add up to 100% (division by the sum of the three threat intensities). Darker colours indicate higher point densities.

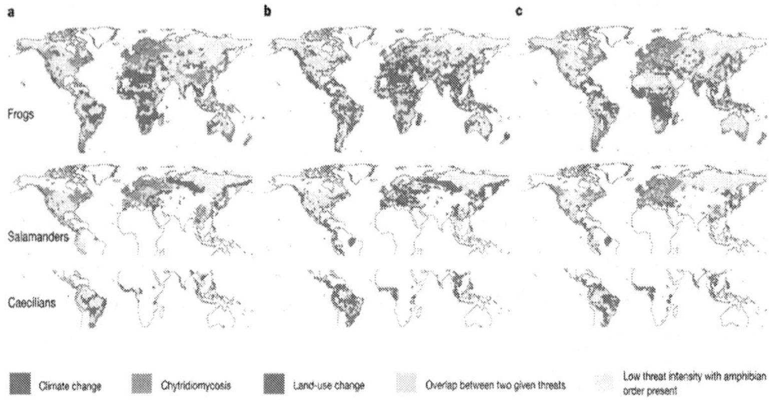

Figure 3: Spatial distribution and pairwise overlap of the three main factors threatening global amphibian biodiversity, projected for the year 2080. a, Climate change and chytridiomycosis. b, Climate change and land-use change. c, Chytridiomycosis and land-use change. Colours indicate areas of particularly high threat intensity. These areas are defined as the 25% of all grid cells with (1) the highest proportion of species projected to lose climatic suitability, (2) the highest projected probability of occurrence of chytridiomycosis and (3) the highest proportion of land projected to be converted from a natural to an anthropogenic state. For other details see Fig. 1.

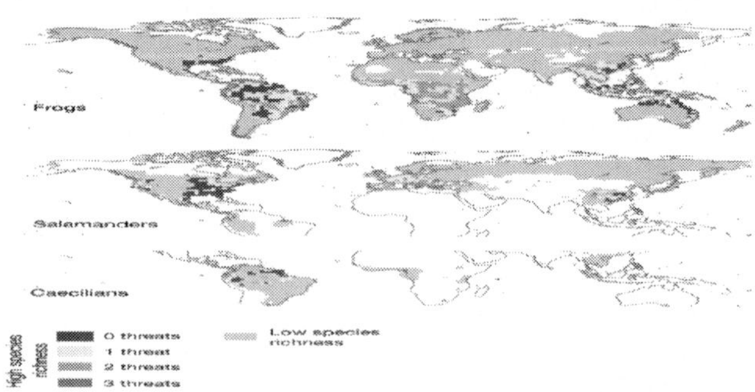

Figure 4: Spatial overlap between areas with the highest amphibian species richness and the main factors threatening global amphibian diversity, pro-

jected for 2080. Areas with the highest species richness are defined as the 25% of all grid cells with the highest number of species. Areas where high levels of species richness coincide with high intensity of 1–3 of the main factors threatening amphibian diversity are coloured in yellow, orange and red, respectively (blue: no coincidence of high richness and high threat intensity). For definition of 25% areas of high threat intensity and further details see Figs 1 and 3.

Several critical assumptions are inevitable when conducting global assessments like ours. Although a careful investigation of some of the methodological uncertainties in our study broadly confirms the consistency of our findings several limitations are important.

First, much uncertainty exists in ensemble forecasting as illustrated by the results produced by our use of 14 different global climate models (GCMs) and three bioclimatic envelope modelling algorithms. Although we explicitly examined this variation, results are contingent on the methods used and future modelling might provide further insights. Second, other threats not investigated here may also cause declines in amphibian populations, including pollution, direct exploitation, spread of invasive species, or increased ultraviolet-B irradiation[5]. These additional factors give reason for further concern, especially when they interact with one another[4]. Third, modelled presence of chytridiomycosis does not always cause local declines or extinctions and interspecific as well as regional differences in amphibian susceptibility to chytridiomycosis exist[22]. Fourth, the spatial resolution of our analysis is relatively coarse, yet several factors allow population persistence at finer scales (see also discussion in refs 23, 24). In particular, our analyses disregard dispersal, biotic interactions and adaptive potential, and rely on the assumption that coarse-scale projections from bioclimatic envelope models provide a good surrogate for species' climatic requirements at finer scales. However, in some cases, species might persist owing to increased climatic stability in fine-grain refugia[25], or to local adaptations of amphibian populations[26]. Our projections thus provide measurements of the exposure of global amphibian species distributions to key threatening processes but, because the species' responses to these threats are not investigated, projections overestimate the impacts of multiple threats on the persistence of local amphibian populations.

Despite these uncertainties, our results reveal an intriguing pattern of non-overlap between key threatening factors. The implication is that risk assessments focusing on a single threat, such as just climate change or only chytridiomycosis, are probably picturing an optimistic view. As such, they fail to identify the key actions required to curb the ongoing global decline in amphibian diversity. The low coincidence between regions projected to have high prevalence of chytridiomycosis and climate or land-use change emphasizes the potential for silent extinctions away from the regions where the current human footprint is larger. In turn, the higher coincidence between land-use and climate change highlights the existence of potential synergies between the two threatening factors that are often neglected[18, 27, 28]. The substantial overlap of threats with many of the world's centres of amphibian richness further underlines the pessimistic long-term perspective for global amphibian diversity[1, 5]. It reinforces the realization that prioritization of conservation efforts needs to be based on knowledge of the spatial distribution both of the different key threats and of biodiversity.

METHODS

Data

We fitted bioclimate envelope models (BEMs) for 5,527 amphibian species from the three amphibian orders Anura (frogs and toads), Caudata (salamanders and newts) and Gymnophiona (caecilians), which for simplicity are henceforth referred to as frogs, salamanders and caecilians. Distribution data were compiled from the Global Amphibian Assessment[31]. Polygons of species' ranges were resampled to a 2 × 2 degree latitude–longitude grid (referred to as the 2°-grid) including 5,041 terrestrial cells (for maps of species richness, see Fig. 1a). This resolution approximates the average of the original resolutions of the climate data sets (see later).

Climatic data were obtained from the World Climate Research Programme (WCRP) Coupled Model Intercomparison Project phase 3 (CMIP3) multi-model data set[30] of the Fourth Intergovernmental Panel on Climate Change (IPCC) report. Data were derived from 14 coupled Atmosphere-Ocean GCMs and 3 emission scenarios. Using this series of GCMs we encompass a wide range of equilibrium climate sensitivity (ECS; 2.1–4.3 °C) and an array of original spatial resolutions, from 1.1 × 1.1 to 3.75 × 3.75 degrees latitude–longitude in the original sets. Outputs for each model were obtained for three SRES[32] emission scenarios A1B, A2 and B1, but A2 and B1 scenarios were not available for all of the GCMs. Inclusion of these three scenarios ensures that the models cover a wide range of likely climatic changes.

For each of the GCMs and emission scenarios, five climatic variables were obtained to characterize the baseline period (averaged across a 30-year time period from 1970 to 1999) that was used to calibrate the models. The same variables were then used to make projections into a 30-year time period between 2070 and 2099, subsequently referred to as 2080. The variables used were mean annual rainfall and precipitation seasonality, annual temperature range, minimum temperature and maximum temperature. These variables are known to impose constraints on amphibian physiology and survival[34] and are often used to model amphibian species distributions and richness[12, 35]. All climate variables were resampled to the 2°-grid.

By using a global extent approach and given the grain of our analysis, we (1) avoid asserting artificial data quality by inappropriate downscaling of the climatic data, and (2) minimize the problem of false absences in the species distribution data[36] as many species in the data set have only been identified in a few localities, with no knowledge about the true occurrence of the species[2]. However, the coarse resolution precludes detailed local assessments of threat interactions and processes; therefore our focus is on documenting coarse spatial patterns.

Projections of the probability of occurrence of *B. dendrobatidis* were obtained from climate-based consensus projections previously described[17]. These projections (standardized probability of occurrence given by a consensus of MaxEnt BEMs across three GCMs and two emission scenarios;) were resampled to the 2°-grid by weighted averaging. In contrast to the climate change and land-use change projections, for chytridiomycosis we did not use values of the change of the probability of occurrence, as many regions are not infected yet by chytridiomycosis, which makes the absolute value of probability of occurrence a better estimate of future risk of *B. dendrobatidis* infection. For the subsequent analyses, we used a consensus map calculated as the arithmetic mean across all the projections (Fig. 1b). Generally, averaging across different scenarios may be problematic. However, for practical reasons and because separate maps of the variation of the probability of chytridiomycosis occurrence for each combination of used GCM × scenario did not show strong differences in the spatial pattern, we used the consensus map.

For the projections of potential land-use change, we used data from the Millennium Ecosystem Assessment[20, 37]. The Millennium Ecosystem Assessment uses four scenarios representing a variety of socio-economic and political futures to estimate future changes in the Earth's land-cover[27] ('Adapting Mosaic', 'Global Orchestration', 'Order from Strength' and 'TechnoGarden'[20]). The Millennium Ecosystem Assessment maps provide information on current and future distributions of 18 different land-cover types at a 0.5° latitude–longitude resolution. For a quantification of potential land-use change we identified grid cells that are projected to change from a natural to an anthropogenic land-cover state (change of any land-cover type to land-cover type 3 'cropland/permanent pasture') and calculated the proportion of area changed for each cell of our 2°-grid for 2080, as a consensus map (arithmetic mean across all four scenarios, Fig. 1c) and separately for each of the four scenarios. As for the projections of chytridiomycosis, for practical reasons and because a separate use of different Millennium Ecosystem Assessment scenarios does virtually not affect the results, we used the consensus map in the subsequent analyses.

Modelling

Three different modelling algorithms, Euclidean distance (ED), Mahalanobis distance (MD) and MaxEnt (MX), were used to run BEMs. These presence-only algorithms were selected owing to the large number of species with uncertain distributions or very small range sizes to be modelled. The two distance-based methods (ED and MD) measure the similarity of each species' occurrence to the mean (or centre) of the available climatic space. Accordingly species' niches are defined as circular (for ED) or elliptical (for MD) shapes in climatic hyperspace[38]. BioEnsembles, a computer software which is able to optimize and take advantage of high-speed parallel processing, was used to run the ED and MD models[39]. MaxEnt version 3.2.4[40, 41], a machine-learning technique based on the principle of maximum entropy, was used to run the MX models. In MaxEnt, we used a regularization multiplier of 0.5 (a model parameter which allows for adjustment of the degree of model overfitting), because this value represents a balance between being able to fit models for species with very few records while avoiding an unreliable degree of overfitting.

For each of the 5,527 species, we ran each possible modelling combination (3 modelling algorithms × 14 GCMs × 3 scenarios × 2 time periods), which resulted in 1,260,156 models (note that for some GCMs only two scenarios were available). Standard BEM validation procedures were not applicable in our study (but note that a validation for future scenarios is in any case not possible[42]). However, we cautiously assessed patterns of variation in model results that may have resulted from different sources of uncertainty, such as species with small numbers of occurrence records, different modelling algorithms, and variation among GCMs that may result from different resolutions and equilibrium climatic sensitivities, as well as different emission scenarios for discussion on model-inherent assumptions). All analyses were performed separately for frogs, salamanders and caecilians.

Processing of Modelling Results

We used a no-dispersal scenario as the basic underlying assumption for the further processing of the modelling results for two reasons. First, it is unlikely that amphibians will be able to fully track changes in climatic conditions by shifts of their distributional ranges[43], in particular when thinking of the coarse spatial scale of our analyses. Second, and more importantly, the ranges of many species are extremely small. Because BEM projections can become unreliable for species with few occurrence records[44, 45, 46], we refrained from projecting a species' range into areas where the species does not currently occur. Furthermore, to avoid uncertainties associated with the choice of thresholds to convert raw model outputs (that is, distance to the optimal centroids or probabilities of occurrence) into binary estimates of presence and absence[47], we decided not to use any type of thresholding. Instead, following ref. 48, we used the change in climatic suitability that overlapped with species' current ranges and disregarded suitability or probability scores that did not overlap with existing records for the species. The change in climatic suitability was then calculated as the difference of the climatic suitabilities between current and future conditions (standardized MX probability of occurrence or 1 minus the raw distance (standardized between 0 and 1) for ED and MD, respectively). This procedure was repeated for each model combination (algorithm × GCMs × scenarios) for each species. Despite the standardization of values of suitability change to a range from 0 to 1, the values are not quantitatively comparable across the different modelling algorithms, which is due to general differences in distance-based (ED, MD) and machine-learning (MX) algorithms as well as to software-inherent differences (for histograms and maps of the mean changes of suitability per grid cell, calculated as the means across all species). Therefore we used a qualitative approach to identify the regions with the strongest projected impacts of climate change on amphibian diversity[48]: for each model combination, we counted the number of species per grid cell that (1) lose climatic suitability ('climate losers': negative change in climatic suitability between current and future conditions), (2) gain climatic suitability ('climate winners': positive change in climatic suitability between current and future conditions) and (3) show no change in climatic suitability between current and future conditions. Note that doing so implies that species will be counted as 'climate loser' or 'climate winner' regardless of the magnitude of those changes, and that species may be identified as losers in one grid cell and as winners in another. However, as we only analyse these counts in categorical and aggregate form (that is, cells with the highest proportion of losers) we expect results to be robust to this simplification and further provide sensitivity analyses.

To identify the regions with the strongest projected impacts of climate change on amphibian diversity, we built consensus maps of the proportion of climate losers and then identified the 25% of all grid cells with the highest proportion of losers (Fig. 3a). Consensus maps were derived by calculating arithmetic means of the proportion of climate losers across all model combinations (algorithm × GCM × scenario) for 2080 (Fig. 1a).

Many studies have shown that species distribution modelling results can vastly differ when using different GCMs, emission scenarios and algorithms[49, 50]. To assess the uncertainties around the consensus, we mapped the proportion of losers separately as arithmetic means (1) across all combinations of GCM × algorithm per scenario, (2) across all combinations of GCM × scenario per algorithm, (3) across all combinations of GCM × scenario per combination of two algorithms, and (4) across all combinations of algorithm × scenario per GCM. Furthermore, following a novel uncertainty assessment protocol[51], we assessed the proportion of variation explained by different sources of uncertainty (algorithm, GCM, scenario, their interactions, and the residual uncertainty) by variance partitioning (SS proportion in 3-way ANOVAs;), and mapped these proportions of uncertainty. In addition, we identified the 25% grid cells with the highest proportion of climate losers separately for each model combination and calculated the number of models per grid cell that identified this grid cell as one of the 25% with the highest proportion of losers. These overlap maps were constructed for each possible algorithm combination (ED × MD × MX; ED × MD, ED × MX, MD × MX; ED, MD, MX) to assess also the amount of uncertainty that is associated with the use of one, two or three modelling algorithms.

Spatial Overlap of Different Threats

To investigate the spatial overlap of different threats for amphibian diversity, we identified the regions with the highest projected impact for each of the respective threat: 25% of all grid cells with the highest proportion of climate losers, 25% of all grid cells with the highest probability of occurrence of *B. dendrobatidis*, and the grid cells with a projected land-use change of at least 25% of the total area. All of these calculations were based on the consensus maps derived using the procedures explained above.

CONTRIBUTIONS

C.H., M.B.A. and C.R. designed the study, C.H. performed all analyses, all authors discussed the results. C.H. wrote the paper, with substantial contributions from all authors.

ACKNOWLEDGEMENTS

We are grateful to D. Rödder, S. Lötters and J. Kielgast for the provision of data and *B. dendrobatidis* modelling results. We thank C. Graham, R. Colwell, N. Sanders, H. H. Bruun and S. Fritz for comments on previous versions of the manuscript. Special thanks to T. Rangel for technical and statistical support. C.H., M.B.A. and C.R. acknowledge the Danish National Research Foundation for support to the Center for Macroecology, Evolution and Climate; research by M.B.A. was funded by the Portuguese Foundation for Science and Technology (PTDC/AAC-AMB/98163/2008); W.J. acknowledges support from NSF grants DBI 0960550 and DEB 1026764.

REFERENCES

1. Stuart, S. N. *et al.* Status and trends of amphibian declines and extinctions worldwide.*Science* 306, 1783–1786 (2004)
2. IUCN. An Analysis of Amphibians on the 2008 IUCN Red List 〈http://www.iucnredlist.org/initiatives/amphibians〉 (2008)
3. Wake, D. B. & Vredenburg, V. T. Are we in the midst of the sixth mass extinction? A view from the world of amphibians. *Proc. Natl Acad. Sci. USA* 105, 11466–11473 (2008)
4. Blaustein, A. R. & Kiesecker, J. M. Complexity in conservation: lessons from the global decline of amphibian populations. *Ecol. Lett.* 5, 597–608 (2002)
5. Beebee, T. J. C. & Griffiths, R. A. The amphibian decline crisis: a watershed for conservation biology? *Biol. Conserv.* 125, 271–285 (2005)
6. Houlahan, J. E., Findlay, C. S., Schmidt, B. R., Meyer, A. H. & Kuzmin, S. L. Quantitative evidence for global amphibian population declines. *Nature* 404, 752–755 (2000)
7. Becker, C. G., Fonseca, C. R., Haddad, C. F. B., Batista, R. F. & Prado, P. I. Habitat split and the global decline of amphibians. *Science* 318, 1775–1777 (2007)
8. Bosch, J., Carrascal, L. M., Duran, L., Walker, S. & Fisher, M. C. Climate change and outbreaks of amphibian chytridiomycosis in a montane area of Central Spain; is there a link? *Proc. R. Soc. Lond. B* 274, 253–260 (2007)
9. Pounds, J. A. *et al.* Widespread amphibian extinctions from epidemic disease driven by global warming. *Nature* 439, 161–167 (2006)
10. Lips, K. R., Diffendorfer, J., Mendelson, J. R. & Sears, M. W. Riding the wave: reconciling the roles of disease and climate change in amphibian declines. *PLoS Biol.* 6, e72 (2008)

11. Lawler, J. J., Shafer, S. L. & Blaustein, A. R. Projected climate impacts for the amphibians of the Western Hemisphere. *Conserv. Biol.* 24, 38–50 (2010)

12. Araújo, M. B., Thuiller, W. & Pearson, R. G. Climate warming and the decline of amphibians and reptiles in Europe. *J. Biogeogr.* 33, 1712–1728 (2006)

13. Bielby, J., Cooper, N., Cunningham, A. A., Garner, T. W. J. & Purvis, A. Predicting susceptibility to future declines in the world's frogs. *Conservation Letters* 1, 82–90 (2008)

14. Sodhi, N. S. *et al.* Measuring the meltdown: drivers of global amphibian extinction and decline. *PLoS One* 3, e1636 (2008)

15. Becker, C. G. & Zamudio, K. R. Tropical amphibian populations experience higher disease risk in natural habitats. *Proc. Natl Acad. Sci. USA* 108, 9893–9898 (2011)

16. Rohr, J. R., Raffel, T. R., Romansic, J. M., McCallum, H. & Hudson, P. J. Evaluating the links between climate, disease spread, and amphibian declines. *Proc. Natl Acad. Sci. USA*105, 17436–17441 (2008)

17. Rödder, D., Kielgast, J. & Lötters, S. Future potential distribution of the emerging amphibian chytrid fungus under anthropogenic climate change. *Dis. Aquat. Organ.* 92, 201–207 (2010)

18. van Vuuren, D. P., Sala, O. E. & Pereira, H. M. The future of vascular plant diversity under four global scenarios. *Ecol. Soc.* 11, 25 (2006)

19. Alcamo, J. *et al.* Global Change Scenarios of the 21st Century, Results from the IMAGE 2.1 Model 3–96 (Elsevier, 1998)

20. Millennium Ecosystem Assessment. Ecosystems and Human Well-Being: Scenarios. (Island, 2005)

21. Daszak, P., Cunningham, A. A. & Hyatt, A. D. Infectious disease and amphibian population declines. *Divers. Distrib.* 9, 141–150 (2003)

22. Kilpatrick, A. M., Briggs, C. J. & Daszak, P. The ecology and impact of chytridiomycosis: an emerging disease of amphibians. *Trends Ecol. Evol.* 25, 109–118 (2010)

23. Hof, C., Rahbek, C. & Araújo, M. B. Phylogenetic signals in the climatic niches of the world's amphibians. *Ecography* 33, 242–250 (2010)

24. Hof, C., Levinsky, I., Araújo, M. B. & Rahbek, R. Rethinking species' ability to cope with rapid climate change. *Glob. Change Biol.* 17, 2987–2990 (2011)

25. Ashcroft, M. B. Identifying refugia from climate change. *J. Biogeogr.* 37, 1407–1413 (2010)

26. Phillimore, A. B., Hadfield, J. D., Jones, O. R. & Smithers, R. J. Differences in spawning date between populations of common frog reveal local adaptation. *Proc. Natl Acad. Sci. USA* 107, 8292–8297 (2010)

27. Jetz, W., Wilcove, D. S. & Dobson, A. P. Projected impacts of climate and land-use change on the global diversity of birds. *PLoS Biol.* 5, e157 (2007)

28. Beaumont, L. J. *et al.* Impacts of climate change on the world's most exceptional ecoregions. *Proc. Natl Acad. Sci. USA* 108, 2306–2311 (2011)

29. Araújo, M. B. & New, M. Ensemble forecasting of species distributions. *Trends Ecol. Evol.*22, 42–47 (2007)

30. Meehl, G. A. *et al.* The WCRP CMIP3 multimodel dataset: a new era in climate change research. *Bull. Am. Meteorol. Soc.* 88, 1383–1394 (2007)

31. IUCN. Global Amphibian Assessment 〈http://www.iucnredlist.org/initiatives/amphibians/〉(2004)

32. IPCC. Special Report on Emissions Scenarios, Prepared for the Third Assessment Report(IPCC, 2000)

33. IPCC. Contribution of Working Group I to the Fourth Assessment Report of the Intergovernmental Panel on Climate Change: Summary for Policymakers (IPCC, 2007)

34. Carey, C. & Alexander, M. A. Climate change and amphibian declines: is there a link?*Divers. Distrib.* 9, 111–121 (2003)

35. Araújo, M. B. *et al.* Quaternary climate changes explain diversity among reptiles and amphibians. *Ecography* 31, 8–15 (2008)

36. Hurlbert, A. H. & Jetz, W. Species richness, hotspots, and the scale dependence of range maps in ecology and conservation. *Proc. Natl Acad. Sci. USA* 104, 13384–13389 (2007)

37. Millennium Ecosystem Assessment. Ecosystems and Human Well-Being: Synthesis(Island, 2005)

38. Farber, O. & Kadmon, R. Assessment of alternative approaches for bioclimatic modeling with special emphasis on the Mahalanobis distance. *Ecol. Modell.* 160, 115–130 (2003)
39. Bioensembles:. software for computer intensive ensemble forecasting of species distributions under climate change v. 1.0 (privately distributed, Madrid, Goiás, Évora, 2009)
40. Phillips, S. J., Anderson, R. P. & Schapire, R. E. Maximum entropy modeling of species geographic distributions. *Ecol. Modell.* 190, 231–259 (2006)
41. Phillips, S. J. & Dudik, M. Modeling of species distributions with Maxent: new extensions and a comprehensive evaluation. *Ecography* 31, 161–175 (2008)
42. Araújo, M. B., Pearson, R. G., Thuiller, W. & Erhard, M. Validation of species–climate impact models under climate change. *Glob. Change Biol.* 11, 1504–1513 (2005)
43. Smith, M. A. & Green, D. M. Dispersal and the metapopulation paradigm in amphibian ecology and conservation: are all amphibian populations metapopulations? *Ecography* 28,110–128 (2005)
44. Stockwell, D. R. B. & Peterson, A. T. Effects of sample size on accuracy of species distribution models. *Ecol. Modell.* 148, 1–13 (2002)
45. McPherson, J. M., Jetz, W. & Rogers, D. J. The effects of species' range sizes on the accuracy of distribution models: ecological phenomenon or statistical artefact? *J. Appl. Ecol.* 41, 811–823 (2004)
46. Wisz, M. S. *et al.* Effects of sample size on the performance of species distribution models.*Divers. Distrib.* 14, 763–773 (2008)
47. Nenzén, H. K. & Araújo, M. B. Choice of threshold alters projections of species range shifts under climate change. *Ecol. Modell.* 222, 3346–3354 (2011)
48. Araújo, M. B., Alagador, D., Cabeza, M., Nogues-Bravo, D. & Thuiller, W. Climate change threatens European conservation areas. *Ecol. Lett.* 14, 484–492 (2011)
49. Araújo, M. B., Whittaker, R. J., Ladle, R. J. & Erhard, M. Reducing uncertainty in projections of extinction risk from climate change. *Glob. Ecol. Biogeogr.* 14, 529–538 (2005)

50. Pearson, R. G. *et al.* Model-based uncertainty in species range prediction. *J. Biogeogr.* 33,1704–1711 (2006)

51. Diniz-Filho, J. A. F. *et al.* Partitioning and mapping uncertainties in ensembles of forecasts of species turnover under climate change. *Ecography* 32, 897–906 (2009)

Chapter 2

Advances in Observation and Estimation of Land Use Impacts on Climate Changes: Improved Data, Upgraded Models, and Case Studies

R. B. Singh[1] and Chenchen Shi[2]

[1]Department of Geography, Delhi School of Economics, University of Delhi, Delhi 110007, India

[2]State Key Laboratory of Water Environment Simulation, School of Environment, Beijing Normal University, Beijing 100875, China

ABSTRACT

Global land use and land cover pattern has greatly changed in the past 50 years, which exerts direct or indirect influence on the climate change remarkably at both regional and global scales. Therefore, observing and estimating the land use impacts on surface climate is essential and has been continuously promoted by researchers. This paper explores the advancement in the models, data, and application for observing and estimating the land use impacts on surface climate and points out further research needs and priorities, which hopefully will provide some references for related studies.

INTRODUCTION

The study of land use and land cover change (LUCC) and its impact on climate is of great significance [1]. There have been tremendous changes in the global land use pattern in the past 50 years, which have enormous influence on the global climate change. Quantitative analysis for the impacts of LUCC on surface climate is one of the core scientific issues to understand the influence of human activities on global climate. For instance, the diverse role of LUCC on precipitation has been documented and land conversion continues at a rapid pace, making this type of human induced disturbances of the climate system continue and become even more significant [2]. The review paper of Deng et al. [1] comprehensively analyzed the primary scientific issues about the impacts of LUCC on the regional climate and reviewed the progress in relevant researches. Earlier as there was no systematic review paper, Deng et al. reviewed the systematic modeling of impacts of LUCC on regional climate for the first time. This paper further develops a systematic review to cover the advances in this research field, in which the improved data, upgraded models, and case studies in observation and estimation of land use impacts on climate changes have been introduced.

ADVANCEMENT IN LUCC DATASET

Previous Dataset

LUCC has been recognized as a key component in global environmental change. It was not until the last twenty years that land cover products were applied to most of the climate models. These products were initially compiled from maps, ground surveys, and various national statistical sources, which have inherent limitations [3–5]. In the mid-1990s, global-scale land cover products generated from remote sensing images became available. Various land cover datasets were usually classified from remote sensing images, including MSS/TM/ETM+, SPOT, and MODIS, which have been used in numerous climate modeling studies at regional to global scales. Since the 1990s, a series of land use/land cover datasets have been produced in many international institutes and countries, such as Global Land Cover Characteristics (GLCC) [6], University of Maryland (UMD) Global Land Cover Facility (GLCF) land cover dataset [7], and Olson Global Ecosystem (OGE) dataset [8, 9]. These products were interpreted artificially with computer aids and then verified by systematic field survey. The need of land use/cover dataset with high accuracy for the climate simulation has been recognized widely in the climate modeling community [10]. Some research showed that the result of the regional precipitation study would be greatly influenced if the accuracy of land cover data is under 80%, and the result may get progressively worse as the accuracy continues to decrease [11]. Therefore the accuracy of the land cover data is of crucial importance to the climate studies. For example, in India and China, large amounts of land cover data have been produced in recent years using the remote sensing data [12]. Unfortunately, neither the overall nor class-specific accuracy of most datasets was able to meet the common requirements of the regional climate modeling.

The disagreement and low accuracy among these land cover datasets primarily resulted from the differences in the sensors, spatial resolutions, algorithms, and classification schemes [13]. Moreover, most of the land cover datasets derived from remote sensing are not 100% accurate. A new statistical measure Q was developed to evaluate land cover datasets in land-climate interaction research, which calculates biophysical precision of land cover datasets using 1 km monthly MODIS Leaf Area Index (LAI) product [14]. The spatial data mining is proposed to produce a higher accuracy land cover map, whose classification system should be compatible with the well accepted classification system used in regional climate simulations [12, 15].

Classification of LUCC Data for Climate Models

There are a number of schemes that have been proposed for regional to global-scale land cover categorization, including the International Geosphere-Biosphere Programme (IGBP)-Data and Information Systems: Land Cover Working Group developed land cover categorization scheme [16], a six-class biome categorization [17], the Simple Biosphere Model scheme [18], and the Federal Geographic Data Committee vegetation characterization and information standards [19]. Many parameters in a land surface model are identified based on the land cover types, for instance, time-invariant model variables (e.g., vegetation reflectance, canopy top height, canopy base height, root depth, and leaf respiration factor) in the Simple Biosphere Model 2 (SIB2) [18] and the Common Land Model (CoLM) [20]. Thus, the specific land cover classification units should not only be discernible (and with high accuracy) from remote sensing image and ancillary data but also be directly related to the physical characteristics of land surface. The IGBP scheme embraces the same philosophy but with modifications to be compatible with existing schemes used by environmental modelers, to incorporate land use in addition to land cover and to represent mosaics [16].

In order to enhance the LUCC study, data need to be updated to increase the accuracy. The temporal land cover datasets have been widely used in numerous climate simulation projects. Most attention has been paid to effects of the accuracy of the land cover data on the climate simulation. Though there are temporal land use data with accuracy higher than 90% [15], still the high-precision land cover data is absent. Therefore, there is an urgent need to reclassify the LUCC dataset to feed into the General Circulation Model (GCM) and Regional Climate Model (RCM). For example, in a case study on North China, Wu et al. [15] overlaid the land cover maps of the IGBP-DIS, Global Land Cover 2000 (GLC), UMD Data, and Data Center for West China (WESTDC), and the compatible grids with classification were selected as the sample grids. They combine the land cover data with the land use data to generate land cover data of high accuracy for the climate simulation. Their study showed that the C4.5 algorithm was suitable for converting the land use data to land cover data of IGBP classification. The temporal land cover data produced by their method can meet the accuracy requirement of the climate simulation and can be used as the parameter of dynamical downscaling in regional climate simulation, which constitutes a significant improvement in the data processing.

Data Resolution and Reliability

Land surface has considerable heterogeneity because of the existence of different land cover types such as bare surface, water bodies, urban land, trees, and snow/ice, which vary over small distance. This surface variability not only determines the microclimate but also affects mesoscale atmospheric circulation [21–23]. The accurate land cover maps are the foundation for land surface, ecological and hydrological modeling, carbon and water cycle studies, and research on global climate change [24, 25]. With many land cover products from different sources becoming available for a given region of the Earth, a challenge arises as to which product is optimal for a land-climate modeling study. Traditional classification accuracy assessment is primarily dependent on ground-based surveys or interpretation of high spatial resolution aerial photos and satellite images. By comparing the classified land cover dataset with the ground-truthing data, error metrics can be developed to report the commission and omission errors. Measures of accuracy, such as the Kappa coefficient of agreement, are frequently calculated to express classification accuracy [26, 27]. Therefore, the researcher should take full advantage of geographic knowledge in the GIS database to support classification, thus to remarkably enhance the classification accuracy of land cover.

Accurate representation of land surfaces is an important factor for climate modeling. However, little attention has been paid to the effect of land cover classification accuracy on climate simulations. In reality, land cover accuracy rarely obtains the commonly recommended 85% target [10]. The accuracy of land use classification can reach 73%–77% using decision-tree classification methods and thereby increasing mapping efficiency by 50% [28]. In addition, most assessments of classified accuracy were conducted using the same dataset as was used to train the classifier. Therefore, the classified accuracy was overstated. Moreover, the spatial data mining techniques for the land cover classification is applied. Finally, the evaluation of the accuracy and uncertainty of the land cover data is accomplished by different methods; the accuracy could reach 88.62% [15]. Inaccurate representation of land cover will lead to differences in simulating sensible heat flux, latent heat flux, and many other variables depending on vegetation and land use parameters. Remote sensing provides accurate representation of Earth's surface on different spatial and temporal scales and is an attractive source for creating high accuracy land cover data. Therefore, it is feasible to take advantage of the existing land cover data from different sources to make a high accuracy land cover data using the spatial data mining method. The information fusion strategy is proposed to produce a higher accuracy land cover map of China [29], whose classification system should be compatible with the well accepted classification system used in land surface modeling.

UPGRADED MODELS

Studies on LUCC processes are often challenged by the complex nature and unexpected behavior of both human drivers and natural constraints. Therefore, a land use change dynamic model was needed to simulate the interdependencies and feedback mechanisms between social, economic, and ecosystem environments. Land use change emerges from the interactions among various components of the coupled human-landscape system, which then feeds back to the subsequent development of those interactions [30]. Currently, there are a lot of evaluation models to simulate the dynamic decision and spatiotemporal process of land use change [1].

Current Models

Empirical Statistical Model

There are numerous empirical statistical models applied to land system dynamics simulation [31, 32]. The empirical statistical model can provide the information of the key driving forces of land use change and reflect the time lag effect of response. Moreover, the data input could be multiscaled. However, certain deficiencies exist in this kind of model, as the model requires to be driven by the data of exogenous land use change rate and amount. Besides, the conversion rule of land use should be manually set. Therefore, the model could not provide references for other regions except for the selected study area.

Econometric Model

The econometric model is a policy evaluation model based on sustainable utilization of land utility which can evaluate the influence of policy factors on land use and promote multifunction sustainable utilization of land.

Sustainability Impact Assessment

Tools for Environmental, Social, and Economic Effects on Multifunctional Land Use in European Regions (SENSOR) is funded by the European Commission FP6 framework research programme to develop tools for ex-ante impact assessment of European policies related to rural land use [33]. It includes a detailed macroeconometric model called NEMESIS, which models cross sector impacts, being the major characteristic of SENSOR project [34]. As it applies a cross-sector approach to land use, it is suitable for large regions. The Dynamics of Land System (DLS) model as another representative econometric model can realize the dynamic simulation of land system on the fine-grid scale through the analysis of land use allocation constraints and simulation of land supply and demand balance. Several case studies show that DLS is able to measure the influence of natural and socioeconomic driving factors to predict the future LUCC, which could provide meaningful decision making for informed land use planning and management [35].

Agent-Based Model (ABM)

Since the 1990s, with the rapid development of computational power, a number of models based on ABM began to be applied in LUCC research, and models based on agent land use/land cover change (ABM/LUCC) have been gradually developed [36]. The agent based model has high flexibility, but the flexibility of the model increases the uncertainty of the model and thus bringing the difficulties for the model validation. The agent based model strives to simulate the individual and group behavior together with decision making model, but the individual decision-making behavior possesses great complexity. Zhang et al. adopted this model in analyzing the impacts of climate change and agents' adaptive behaviors on the regional land use changes [37]. In addition, their research showed that the model is sensitive to the climatic conditions, which gave credit to ABM on modeling the LUCC impact on climate.

Coupled Econometric and ABM Model

The combination of agent based model and the econometric model can simulate complex decision problem of land use in better way, so as to provide new methods for simulation of land use dynamics.

The agent model and the econometric model of coupling in the areas of land use will get recognition and application in the future, mainly embodied in the following two aspects: the combination of the agent based model and the econometric model can simulate the individual and group decision behavior considering the interaction among the micro- and macrodecisions at the same time.

The integration of the economic and agent based model combines the process of social and economic factors with the land use dynamics. The comprehensive model not only can be used in the simple scenario simulation of land use but also can be used in complex scenario. The two models, interconnected and interdependent, constitute a complex system with hierarchical heterogeneity.

Among recent studies on land use models, Deng et al. introduced a land use change dynamic (LUCD) model embedded in RCM [38]. The Agro-ecological Zone (AEZ) model is intended as the optimal option for constructing one of the constituent modules, which is naturally correlated with AEZs. The agent-based module identifies the land use change demand and vegetation change can be realized and provides the land use change simulation results which are the underlying surfaces needed by RCM. By importing the RCM's simulation results of climate change and providing the simulation results of land use change for RCM, the LUCD model can be embedded in RCM. The coupled simulation system of LUCD and RCM will be extraordinarily powerful in land surface system simulation. The innovation of their study lies in that current research could only simulate the land cover change through vegetation change modeled in RCM and GCM but not the land use change. But this study simulated dynamic change of land use, which is a great breakthrough in systematically reporting the land use change.

Scenario Development

Scenario analysis of land use change and dynamic prediction involves many driving factors that affect land use change, which has always been the attention of academia worldwide. Through simulating regional land use change under different scenarios, rational forecasting and evaluation of the future land use could be made, which will offer scientific reference for regional and national scale land use planning and management.

Climate scenarios are plausible representations of future climate conditions, which can be produced using a variety of approaches, and among which, regional climate models are increasingly popular [39]. There is a notable increase in interest in regional-scale climate scenarios and projection methods, especially for impact and adaptation assessment [40].

The Intergovernmental Panel on Climate Change (IPCC) has been working on the scenarios of potential future anthropogenic climate change, the underlying driving forces, and the response options. According to IPCC [41, 42], socioeconomic scenarios could facilitate the exploration of long term consequences of anthropogenic climate change and available response options. The socioeconomic futures in the scenario consist of qualitative narrative descriptions of future trends and quantitative assumptions (also called the storyline) about key socioeconomic variables.

In the study of land use impact on climate, with a good database, an appropriate method shall be adopted to facilitate the study, and the scenario analysis can be adopted here, which performs macrostructure change. In the study of Xu et al. [43], three kinds of scenarios were designed based on the socioeconomic development. They simulated land use trends under three different scenarios including baseline scenario and two control scenarios (risk scenario and rapid development mode transformation scenario) and obtained the most suitable scenario to control the CO_2 emissions of the three scenarios. Their scenarios excel other scenarios as they compile the IPCC scenario with the specific situation in the research area. With the scenario analysis, it will be possible to provide a theoretical basis for the future land use planning in adaptation to climate change. As the first attempt of the systematic analysis of LUCC scenarios and with the dataset produced by their study, their research laid a good foundation for future researches.

LAND USE IMPACTS ON CLIMATE CHANGES: CASE STUDIES

With an improved database and models, the climate effect model shall be introduced in order to apply the simulation platform of the land use impact on the climate. Climate change simulation has increased in many countries like China during recent years. Many scholars offered representative methods; among them, Gao et al. simulated the climate change in China using the Abdus Salam International Centre for Theoretical Physics (ICTP) Regional Climate Model version 3 (RegCM3) [44]. The RegCM3 can reproduce the observed spatial structure of surface air temperature and precipitation. However, so far RegCM3 is a hydrostatic model which limits the further increase of its resolution beyond ~10 km [45]. Wu and Fu used a linear fitting method to estimate the change of precipitation intensity spectra at different spatial scales in China [46].

As a study on climate effect, the Weather Research and Forecasting (WRF) Model could be correlated. Qu et al. [47] used WRF to explore the impacts of land cover change on the near-surface temperature in a research in Northern Plain in China. They tested and verified the ability of the WRF to simulate the near-surface temperature in the North China Plain. Notably, in this study the LUCC dynamic model and LUCC reclassification were also applied to form a good systematic modeling of LUCC. Coupled with WRF Model, land use impact on climate was well simulated.

Based on the compiled climate model, the climate change influence on climate index could be further explored. Zhang et al. [48] first reported the climate impact on grassland. Their study took a macro and mesoscale analysis of the possible changing trends of the net primary productivity (NPP) of local grasslands under four RCPs scenarios. The results showed that the grassland productivity will be greatly influenced by the fluctuations of precipitation and temperature.

With the above mentioned studies, LUCC and future climatic scenarios simulation in particular areas were conducted. Simulation of LUCC and its climatic effects were integrated through linking the functions of the newly developed models with those of the existing models.

Further, studies related to land use changes associated with climate variations were promoted with the advances in the research methods. Deng et al. [49] conducted a revisit to the impact of land use changes on the human well-being via altering the ecosystem provisioning services. LUCC and climate change exert tremendous influence on ecosystem provisioning services by changing agro-ecosystem services functions and forest and/or grassland ecosystem services functions. In order to tackle the ecosystem provisioning services problems, measures were taken, which increased the input and reduced the outcome, at the same time augmented the health risk and harm to human well-being.

R. B. Singh and S. Singh [50] have employed pixel based high resolution mapping and modeling at small scale land use pattern and land cover analysis to identify induced flood risks in the Yamuna and the Hindon basin area by WetSPA Model. Flood forecasting has been done with the help of ILWIS Model while LUCC has been analyzed with the help of SRTM, IKNOS, and Landsat data to find out that almost 36% forest land has been transformed into settlement and other land uses. Yamashima et al. [51] have analyzed the impact of historical LUCC on the seasonal cycle of hydroclimate over the Indian subcontinent and southern China. They investigated using a general circulation model MIROC3.2 coupling with land surface model MATSIRO over the period 1700–1850. A decrease in evapotranspiration has been observed over the Indian subcontinent due to decrease in precipitation which is induced by changes in land use/ cover.

Sahu et al. [52] have investigated the impact of recently invented global climatic variability modes like Indian Ocean Dipole (IOD) and El Niño Modoki at the regional climate scale particularly at the river catchment level on daily time series and concluded that at the upper river catchment climate variability has significant impact on seasonal stream flows and land use pattern. Sen Roy et al. [53] also conducted a comprehensive study to understand the spatial pattern of Urban Heat Islands (UHI) development within the rapidly changing urban landscape of Delhi metropolitan region and Kolkata metropolitan region [54]. In both the studies land use has been a key driving force for change in microclimate. In Delhi, overall higher temperatures were observed in the northern and western section, which constitute the most densely built-up residential areas and thereby experiencing rapid vegetation loss and land use change. Remote sensing and GIS based tools have been effectively used to predict UHI risks.

Recently, satellite meteorology products and applications get more and more importance. However, it cannot replace surface measurements. For example, satellite data in the form of GCMs needs observed data for model validation and improves the quality of statistical and scientific significances in research methods. Therefore, increasing emphasis should be given to integrate observed and simulated datasets for improving understanding about land use impacts on climate change [55, 56].

CONCLUDING REMARKS

This paper reviewed the progress of the research in observing and estimating the land use impacts on surface climate from the perspective of the improved models, enhanced data, and widely expanded applications. The major findings include the following: the reclassification of the temporal land cover data can meet the accuracy requirement of the climate simulation and can be used as the parameters of dynamical downscaling in regional climate simulation, which enhances the data processing and lays a good foundation for LUCC simulation. The Land Use Change Dynamics Model compatible with RCMs is a great breakthrough in systematically reporting the dynamic change of land use. The scenario analysis performed the macrostructure change, which facilitated the study of land use impact on climate. LUCC dynamic model and LUCC reclassification can be applied to form a good systematic modeling of LUCC coupled with the WRF Model; land use impact on climate was well simulated.

Despite the progresses in the research, there are still some research needs, which should be further addressed. For example, the LUCC parameters need to be precisely expressed to make sure of the accuracy of the simulation. It is necessary to strengthen the observation of the land surface processes due to the variations in the impact of land use change on climate.

The researches observing and estimating land use impact on climate should mainly focus on two aspects [1]: (i) exploring the climate effects of the future LUCC as the current study has mainly been addressed on the past and current LUCC and (ii) predicting the land use changes in the future and assessing its influence on the future climate.

ACKNOWLEDGMENTS

This research was financially supported by the National Key Programme for Developing Basic Science in China (Grant no. 2010CB950900), the major research plan of the National Natural Science Foundation of China (Grant no. 91325302), and the National Natural Science Funds of China for Distinguished Young Scholar (Grant no. 71225005).

REFERENCES

1. X. Deng, C. Zhao, and H. Yan, "Systematic modeling of impacts of land use and land cover changes on regional climate: a review," Advances in Meteorology, vol. 2013, Article ID 317678, 11 pages, 2013. ·
2. R. A. Pielke, J. Adegoke, A. Beltrán-Przekurat et al., "An overview of regional land-use and land-cover impacts on rainfall," Tellus, Series B, vol. 59, no. 3, pp. 587–601, 2007. · ·
3. E. Matthews, "Global vegetation and land use: new high-resolution data bases for climate studies,"Journal of Climate & Applied Meteorology, vol. 22, no. 3, pp. 474–487, 1983.
4. J. S. Olson, J. A. Watts, and L. J. Allison, "Carbon in live vegetation of major word ecosystems," Tech. Rep. ORNL-5862, Oak Ridge National Laboratory, Oak Ridge, Tenn, USA, 1983.
5. J. Cihlar, "Land cover mapping of large areas from satellites: status and research priorities,"International Journal of Remote Sensing, vol. 21, no. 6-7, pp. 1093–1114, 2000.
6. T. R. Loveland, B. C. Reed, J. F. Brown et al., "Development of a global land cover characteristics database and IGBP DISCover from 1 km AVHRR data," International Journal of Remote Sensing, vol. 21, no. 6-7, pp. 1303–1330, 2000.
7. M. C. Hansen and B. Reed, "A comparison of the IGBP DISCover and University of Maryland 1 km global land cover products,"

International Journal of Remote Sensing, vol. 21, no. 6-7, pp. 1365–1373, 2000.

8. J. Olson, "Global ecosystem framework-definitions," USGS EROS Data Center Internal Report, Sioux Falls, SD, vol. 37, 1994.
9. J. Olson, "Global ecosystem framework-translation strategy," USGS EROS Data Center Internal Report, Sioux Falls, SD, vol. 39, 1994.
10. J. Ge, J. Qi, B. M. Lofgren, N. Moore, N. Torbick, and J. M. Olson, "Impacts of land use/cover classification accuracy on regional climate simulations," Journal of Geophysical Research D, vol. 112, no. 5, Article ID D05107, 2007. · ·
11. J. J. Feddema, K. W. Oleson, G. B. Bonan et al., "Atmospheric science: the importance of land-cover change in simulating future climates," Science, vol. 310, no. 5754, pp. 1674–1678, 2005. · ·
12. Y. H. Ran, X. Li, L. Lu, and Z. Y. Li, "Large-scale land cover mapping with the integration of multi-source information based on the Dempster-Shafer theory," International Journal of Geographical Information Science, vol. 26, no. 1, pp. 169–191, 2012. · ·
13. M. Herold, P. Mayaux, C. E. Woodcock, A. Baccini, and C. Schmullius, "Some challenges in global land cover mapping: an assessment of agreement and accuracy in existing 1 km datasets," Remote Sensing of Environment, vol. 112, no. 5, pp. 2538–2556, 2008. · ·
14. J. Ge, N. Torbick, and J. Qi, "Biophysical evaluation of land-cover products for land-climate modeling,"Earth Interactions, vol. 13, no. 6, 2009. · ·
15. F. Wu, J. Zhan, H. Yan, C. Shi, and J. Huang, "Land cover mapping based on multisource spatial data mining approach for climate simulation: a case study in the farming-pastoral ecotone of North China,"Advances in Meteorology, vol. 2013, Article ID 520803, 12 pages, 2013. ·
16. A. Belward and T. Loveland, "The IGBP DIS 1 km land cover project: remote sensing in action," inProceedings of the 21st Annual Conference of the Remote Sensing Society, pp. 1099–1106, Southampton, UK, 1995.

17. S. W. Running, T. R. Loveland, and L. L. Pierce, "A vegetation classification logic based on remote sensing for use in global biogeochemical models," Ambio, vol. 23, no. 1, pp. 77–81, 1994.
18. P. J. Sellers, S. O. Los, C. J. Tucker et al., "A revised land surface parameterization (SiB2) for atmospheric GCMs. Part II: the generation of global fields of terrestrial biophysical parameters from satellite data,"Journal of Climate, vol. 9, no. 4, pp. 706–737, 1996.
19. F. G. D. Committee, "Vegetation Subcommittee. 1997," FGDC Vegetation Classification and Information Standards, vol. 3, 1996.
20. X. Dai, X. Zeng, and R. Dickinson, "The Common Land Model (CLM): technical documentation and user›s guide," Georgia Institute of Technology, 2001.
21. D. L. Hartmann, Global Physical Climatology, Academic Press, 1994.
22. C. P. Weaver and R. Avissar, "Atmospheric disturbances caused by human modification of the landscape," Bulletin of the American Meteorological Society, vol. 82, no. 2, pp. 269–281, 2001.
23. Z. L. Yang, Modeling Land Surface Processes in Short-Term Weather and Climate Studies, World Scientific Series on Meteorology of East Asia, 2004.
24. I. G. Change, Report, 1990.
25. P. J. Sellers, R. E. Dickinson, D. A. Randall et al., "Modeling the exchanges of energy, water, and carbon between continents and the atmosphere," Science, vol. 275, no. 5299, pp. 502–509, 1997. · ·
26. R. G. Congalton, "A review of assessing the accuracy of classifications of remotely sensed data," Remote Sensing of Environment, vol. 37, no. 1, pp. 35–46, 1991.
27. G. M. Foody, "Status of land cover classification accuracy assessment," Remote Sensing of Environment, vol. 80, no. 1, pp. 185–201, 2002. · ·
28. C. Homer, C. Huang, L. Yang, B. Wylie, and M. Coan, "Development of a 2001 national land-cover database for the United States," Photogrammetric Engineering and Remote Sensing, vol. 70, no. 7, pp. 829–840, 2004.

29. Y. Ran, X. Li, and L. Lu, "Evaluation of four remote sensing based land cover products over China,"International Journal of Remote Sensing, vol. 31, no. 2, pp. 391–401, 2010. · ·
30. Q. B. Le, S. J. Park, P. L. G. Vlek, and A. B. Cremers, "Land-Use Dynamic Simulator (LUDAS): a multi-agent system model for simulating spatio-temporal dynamics of coupled human-landscape system. I. Structure and theoretical specification," Ecological Informatics, vol. 3, no. 2, pp. 135–153, 2008. · ·
31. P. H. Verburg, B. Eickhout, and H. Meijl, "A multi-scale, multi-model approach for analyzing the future dynamics of European land use," The Annals of Regional Science, vol. 42, no. 1, pp. 57–77, 2008. · ·
32. J. Y. Liu and X. Z. Deng, "Progress of the research methodologies on the temporal and spatial process of LUCC," Chinese Science Bulletin, vol. 55, no. 14, pp. 1354–1362, 2010. · ·
33. K. Helming, B. Konig, and K. Tscherning, "SENSOR first annual report (public part)," Management, vol. 30, no. 3, pp. 391–405, 2006.
34. T. Jansson, M. Bakker, B. Boitier et al., "Cross sector land use modelling framework," in Sustainability Impact Assessment of Land Use Changes, pp. 159–180, Springer, 2008.
35. X. Deng, Q. Jiang, J. Zhan, S. He, and Y. Lin, "Simulation on the dynamics of forest area changes in Northeast China," Journal of Geographical Sciences, vol. 20, no. 4, pp. 495–509, 2010. · ·
36. T. P. Evans and H. Kelley, "Multi-scale analysis of a household level agent-based model of landcover change," Journal of Environmental Management, vol. 72, no. 1-2, pp. 57–72, 2004. · ·
37. T. Zhang, J. Zhan, J. Huang, R. Yu, and C. Shi, "An Agent-based reasoning of impacts of regional climate changes on land use changes in the Three-River Headwaters Region of China," Advances in Meteorology, vol. 2013, Article ID 248194, 9 pages, 2013. ·
38. X. Deng, J. Liu, Y. Lin, and C. Shi, "A framework for the land use change dynamics model compatible with RCMs," Advances in Meteorology, vol. 2013, Article ID 658941, 7 pages, 2013. ·

39. R. H. Moss, J. A. Edmonds, K. A. Hibbard et al., "The next generation of scenarios for climate change research and assessment," Nature, vol. 463, no. 7282, pp. 747–756, 2010. · ·
40. S. Solomon, D. Qin, M. Manning et al., "The physical science basis," in Contribution of Working Group I to the Fourth Assessment Report of the Intergovernmental Panel on Climate Change, pp. 235–337, 2007.
41. E. Kriegler, B. O›neill, S. Hallegatte et al., CIRED Working Paper DT/WP, 2010.
42. N. Nakicenovic, J. Alcamo, G. Davis et al., Special Report on Emissions Scenarios: a special report of Working Group III of the Intergovernmental Panel on Climate Change, Pacific Northwest National Laboratory, Environmental Molecular Sciences Laboratory (US), Richland, Wash, USA, 2000.
43. Q. Xu, Q. Jiang, K. Cao, X. Li, and X. Deng, "Scenario-based analysis on the structural change of land uses in China," Advances in Meteorology, vol. 2013, Article ID 919013, 12 pages, 2013. ·
44. X. Gao, Y. Shi, D. Zhang, and F. Giorgi, "Climate change in China in the 21st century as simulated by a high resolution regional climate model," Chinese Science Bulletin, vol. 57, no. 10, pp. 1188–1195, 2012. · ·
45. J. Wu, X. Gao, F. Giorgi, Z. Chen, and D. Yu, "Climate effects of the three gorges reservoir as simulated by a high resolution double nested regional climate model," Quaternary International, vol. 282, pp. 27–36, 2012.
46. F. Wu and C. Fu, "Change of precipitation intensity spectra at different spatial scales under warming conditions," Chinese Science Bulletin, pp. 1–10, 2013.
47. R. Qu, X. Cui, H. Yan, E. Ma, and J. Zhan, "Impacts of land cover change on the near-surface temperature in the North China Plain," Advances in Meteorology, vol. 2013, Article ID 409302, 12 pages, 2013. ·
48. R. Zhang, Z. Li, Y. Yuan, Z. Li, and F. Yin, "Analyses on the changes of grazing capacity in the Three-River Headwaters Region of China under various climate change scenarios," Advances in Meteorology, vol. 2013, Article ID 951261, 9 pages, 2013. ·

49. X. Deng, Z. Li, J. Huang, Q. Shi, and Y. Li, "A revisit to the impact of land use changes on the human well-being via altering the ecosystem provisioning services," Advances in Meteorology, vol. 2013, Article ID 907367, 8 pages, 2013.
50. R. B. Singh and S. Singh, "Rapid urbanization and induced flood risk in Noida, India," Asian Geographer, vol. 28, no. 2, pp. 147–169, 2011.
51. R. Yamashima, K. Takata, J. Matsumoto, and T. Yasunari, "Numerical study of the impacts of land use/cover changes between 1700 and 1850 on the seasonal hydroclimate in monsoon Asia," Journal of the Meteorological Society of Japan, vol. 89, pp. 291–298, 2011. · ·
52. N. Sahu, S. K. Behera, Y. Yamashiki, K. Takara, and T. Yamagata, "IOD and ENSO impacts on the extreme stream-flows of Citarum river in Indonesia," Climate Dynamics, vol. 39, no. 7-8, pp. 1673–1680, 2012. · ·
53. S. Sen Roy, R. B. Singh, and M. Kumar, "An analysis of local spatial temperature patterns in the Delhi Metropolitan Area," Physical Geography, vol. 32, no. 2, pp. 114–138, 2011. · ·
54. R. B. Singh and R. Nath, "Remote sensing, GIS and micrometeorology for monitoring and predicting urban heat islands in Kolkata mega city," Annals of NAGI, vol. 32, no. 1, pp. 17–39, 2012.
55. R. B. Singh, J. Fox, and Y. Himiyama, Eds., Land Use and Cover Change, Science Publishers, Enfield, Conn, USA, 2001.
56. Y. Himiyama and I. Bicik, Eds., Land Use Changes in Selected Regions in the World, Hokkaido University of Education, 2012.

Chapter 3

Effect of Land Use and Climate Change on Runoff in the Dongjiang Basin of South China

Yanhu He[1, 2] Kairong Lin[1, 2], and Xiaohong Chen[1, 2]

[1]Center for Water Resources and Environment, Sun Yat-Sen University, Guangzhou 510275, China

[2]Key Laboratory of Water Cycle and Water Security in Southern China of Guangdong Higher Education Institutes, Guangzhou 510275, China

ABSTRACT

Variability and availability of water resources under changing environment in a regional scale have been hot topics in recent years, due to the vulnerability of water resources associated with social and economic development. In this paper, four subbasins in the Dongjiang basin with a significant land use change were selected as case study. Runoffs of the four subbasins were simulated using the SCS monthly model to identify the quantitative impacts of land use and climate change. The results showed that (1), in the Dongjiang basin, temperature increased significantly, evaporation and sunlight decreased strongly, while precipitation showed a nonsignificant increase; (2) since the 1980s, land uses in the Dongjiang basin have experienced a significant change with a prominent increase in urban areas, a moderate increase in farmlands, and a great decrease in forest areas; (3) the SCS monthly model performed well in the four subbasins giving that the more significant land use change in each subbasin, the more runoff change correspondingly; (4) overall, runoff change was contributed half and half by climate change and human activities, respectively, in all the subbasins, in which about 20%~30% change was contributed by land use change.

INTRODUCTION

Environment Change (including land use change and climate change) and its impacts on water resources have always been the hot issues in recent years. The direct or indirect impacts on the hydrological regime brought by land use and climate change both have contributed to some water problems, such as water shortage, flooding, and water logging to different extent. Some researchers have been conducted to study the impacts of land use and climate change on water resources in different basins [1–5]. Particularly in the humid region of south China, six hydrological models were used to simulate the hydrological impact of some climate change scenarios, and response strategies for water supply and flood control due to climate change were analyzed in the Dongjiang basin by Jiang et al. [6]. South China has identified an increasing trend of extreme rainfall events, and the correlation between such events and flood events was studied by Fu et al. [7]. Climate change makes flooding more frequent in some regions [8, 9]. Additionally, land use change also has notable impact on the hydrological cycle [10–12]. Overall, hydrological process and the variability and availability of water resources would change a lot due to land use change and climate change.

As the biggest developing country in the world, China has experienced an explosive economic growth over a couple of decades, which results in the prominent land use change. This change can alter evaporation patterns and potentially affect water resources in a region. The impacts of climate change on water resources and agriculture in

China were analyzed by Piao et al. [13], and the results indicated that agriculture relying on the resultant increase in glacier runoff, especially in western China, would face a challenge, although its climate trends remained moderate compared to natural variability [13]. Even worse was the more frequent occurrence of drought and flood extremes largely due to climate change in different parts of China [14]. Some researchers have focused on the variability of runoff in the Dongjiang basin under the impacts of climate change and human activities, especially land use change in recent decades [15]. Definitely, there is an increasingly serious challenge in the availability of water resources, and its exploitation and management will be more or less affected.

As to a basin, climate change and human activities both contribute to the hydrological cycle, and this finding has been supported by many studies [16–19]. Many studies have been carried out to distinguish the roles of land use and climate change on water resources. The approach to the calculation for the runoff change due to precipitation and potential evaporation was proposed [20–22]. Based on which, the separated impacts of human activities and climate change on natural runoff change in the Poyang basin from 1992 to 2000 were analyzed quantitatively by Ye et al. [23]. In recent years, much attention has been paid to analyze the separated quantitative effects of land use and climate change, and some progress has been made, although further research still needs to be carried out. A simple approach to distinguish land-use and climate-change effects on water resources was developed with a coupled water-energy budget analysis in US Midwest area by Tomer and Schilling [11]. Combined impacts of land cover and climate changes on hydrological processes of the Kejie watershed in the eastern Himalayas were assessed with a SWAT model, which revealed that land-cover change had more effects than those of climate change in the short and middle terms [24]. Furthermore, in many river basins of China, many studies have been conducted to assess the impacts of climate change and human activities on the runoff variation. As to Tarim River, the impact of human activities had increased the runoff with a ratio of 41%~75% for different ages [25], and similar research can be found in the Wuding River [26]. In addition, some relative studies also took place in the Heihe catchment, Chaobai River, and Mian River basin [27–29], which came to a conclusion that climate change and human activities have the separated impacts on the runoff change varying from place to place. There is no exception for the Dongjiang basin

[30]. However, few researches have been conducted to identify the quantitative effects of human activities and climate change on runoff in the study area. Moreover, most of the previous studies focused on a large scale basin and had no cases for comparison, which tends to limit our understanding of the impacts of climate change and human activities on water resources in the study area. The quantitative effect of land use change on runoff of the basin has not been revealed yet. Therefore, it is desirable to separate the impacts of climate change and human activities, specifically the land use change, on water resources under a changing environment in the study area.

Hydrological models have been widely used to study many practical and pressing issues that arise during planning, design, operation, and management of water resources systems [31, 32] and also to quantify the impacts of land use and climate change on the hydrological cycle. For example, a conceptual rainfall-runoff model was applied to the Rhine basin for the purpose of modeling the effect of land use change on the runoff. The results suggested that increased urbanization led to an increase in the lower peak runoff, compared to a considerable reduction of both the peak runoff and the total runoff volume resulted from intensified forestation [33]. Besides, the hydrological model considering other elements such as biogeochemistry was applied to reveal how land use change affects hydrological regimes at the watershed scale [34]. According to the available climate and hydrological data and the hydrological characteristics of the basin studied, many conceptual or distributed hydrological models were introduced to analyze the hydrological response under land use and climate change [35–38]. Trend analysis of climate variables is necessary to detect the variability of climate variables, such as temperature and precipitation [39, 40], which provides supports for study of the impacts on the hydrological cycle of land use change and climate change. Overall, hydrological model combined with statistic methods has been a prevalent and useful tool to clarify such an interesting phenomenon for a long time.

The Dongjiang River, which lies in south China, is an important fresh water source of the Pearl River Delta (one of the most developed areas in China). Specifically, it supplies 70% fresh water for Hong Kong. However, water shortage and water pollution in the Dongjiang basin appear to be more and more serious in recent years, due to the fast and persistent economic development and urbanization. Land use in this region has changed prominently since 1980s and contributed

Organization [43], the Mann-Kendall test was widely used to detect time-series trends in hydrological and climatic data in many researches. The Mann-Kendall test has some advantages including its simplicity, ability to deal with non-normal and missing data distributions, and robustness to the effects of outliers and gross data errors [44–46].

Firstly, the Mann-Kendall test defined a variable S as:

$$S = \sum_{k=1}^{n-1} \sum_{j=k+1}^{n} \operatorname{sgn}\left(x_i - x_j\right). \tag{1}$$

Then the presence of a statistically significant trend is evaluated using the Z value:

$$Z = \begin{cases} \dfrac{S-1}{\sqrt{\operatorname{Var}(S)}}, & S > 0, \\ 0, & S = 0, \\ \dfrac{S+1}{\sqrt{\operatorname{Var}(S)}}, & S < 0. \end{cases} \tag{2}$$

Besides, the standard normal cumulative distribution function F_n under the significance level α is given by $F_n\,(Z_\alpha/2) = \alpha/2$. If $|Z| > Z_\alpha/2$, the hypothesis of H_1 for a two-sided test can be accepted. Positive value Z indicates an upward trend while negative value a downward trend. More descriptions of this method can be found in many researches [47–49]. In this paper, trend analysis of climatic variables and runoff in the Dongjiang basin were conducted with the Mann-Kendall test based on over 50 years of climatic and hydrological data.

In addition, mutation detections of the climatic data were also conducted by the Mann-Kendall mutation analysis test method; a statistic variable of this method is

$$S_K = \sum_{i=1}^{n} r_i, \quad k = 2, 3, \ldots, n,$$

$$r_i = \begin{cases} +1, & x_i > x_j, \\ 0, & x_i \leq x_j, \end{cases} \quad j = 1, 2, \ldots, \tag{3}$$

Where S_K is the counts for the x series when x_i is greater than x_j

UF_K is the standard normal distribution and can be calculated based on the x series:

$$UF_K = \frac{[S_K - E(S_K)]}{\sqrt{\operatorname{var}(S_K)}}, \quad k = 1, 2 \ldots \tag{4}$$

Where $UF_1 = 0$, $E(S_K)$ and var (S_K) are the expected value and variance of S_K which can be calculated on condition that the x series is mutually independent with the same continuous distribution:

$$E(S_K) = \frac{n(n+1)}{4},$$

$$\operatorname{var}(S_K) = \frac{n(n-1)(2n+5)}{72}. \tag{5}$$

SCS Rainfall-Runoff Model

Developed by the Soil Conservation Service of U.S. Department of Agriculture (USDA) early in 1954, the SCS model is widely used in the USA and many other countries [50, 51]. Several advantages of this model are as follows (1) the characteristics of the land cover such as soil, slope, vegetation and land use in a basin can be considered with the SCS model; the possible change on the rainfall-runoff relationship can be preliminarily estimated according to the change of land use, and (2) it has the advantages of simplicity in structure and convenience in use with very low dependence on data.

The runoff yield for the SCS model is

$$R = \begin{cases} \frac{(P - I_a)^2}{P + S - I_a}, & P \geq I_a, \\ 0, & P < I_a, \end{cases} \tag{6}$$

Where R is the runoff, P is the precipitation, I_a is the initial loss, and S is the present probable maximum retention in the basin which is the top limit of the later loss.

Initial loss I_a can be calculated by an empirical correlation with S:

$$I_a = \alpha S, \tag{7}$$

Where α is the coefficient of the initial losses which is related to the initial soil water content of the basin

To eliminate the large variation scope of the S value, there is an empirical relation for CN (a non-dimensional parameter) and S:

$$S = \frac{25400}{CN} - 254. \tag{8}$$

The curve number CN is a key and comprehensive parameter within SCS model. CN describes the watershed features before a rain and is affected by AMC (antecedent moisture condition), slope, vegetation, soil type, and land use condition with a value of 0~100. AMC can be divided into 3 classes: AMC I-arid condition, AMC II-normal condition, and AMC III-moist condition. The specific basis for the classification and the estimation of the CN value can be found in related references [52].

Input by the observed monthly precipitation and evaporation, the SCS model yields the monthly runoff. A key issue here is to calculate the actual evaporation by using the SCS monthly model. Based on the research results from the two-parameter monthly water balance model [53], the monthly actual evaporation can be calculated by

$$E(t) = C \times EP(t) \times \tanh\left(\frac{P(t)}{EP(t)}\right), \tag{9}$$

Where (t) is the actual monthly evaporation, $EP(t)$ is the observed evaporation from the evaporating dish, $P(t)$ is the monthly precipitation, and C is one of the parameters in the model (non-dimensional). The total discharge consists of two parts: the surface flow ($S(t)$) and the base flow ($RG(t)$). $S(t)$ can be gained by (6), while $RG(t)$ can be calculated with

$$RG(t) = a \times Z(t), \tag{10}$$

Where a is the coefficient of the groundwater flow and (t) is the soil moisture content which can be calculated by

$$Z(t) = Z(t-1) + P(t) - E(t) + SR(t) - RG(t). \tag{11}$$

A schematic diagram for the water balance in a basin is shown in Figure 5. The SCS model is used to simulate the monthly runoff process in this study. There are two parameters, C and α, in the model

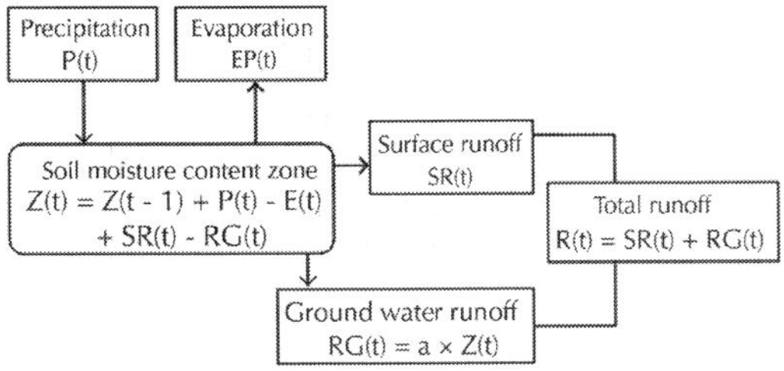

Figure 5: A schematic diagram of the water balance in a basin.

Approach to Distinguish the Respective Impact of Land Use and Climate Change to Water Resources

The total runoff change can be obtained by the difference between the observed runoffs in two different periods, respectively, that is, the intensive human activities period and the low human activities period (natural condition). It is assumed that the runoff change in the low human activities period is affected only by climate change.

Therefore, the impacts of climate change, land use change, and other human activities on runoff change can be separated according to the following calculations:

$$\Delta R_L = R_{H2} - R_{H1},$$

$$\Delta R_C = R_B - R_{H1},$$

$$\Delta R_0 = \Delta R_T - \Delta R_C - \Delta R_L,$$

$$\Delta = |\Delta R_0| + |\Delta R_C| + |\Delta R_L|,$$

$$\mu L = \frac{|\Delta R_L|}{\Delta} \times 100\%,$$

$$\mu C = \frac{|\Delta R_C|}{\Delta} \times 100\%,$$

$$\mu R_0 = \frac{|\Delta R_0|}{\Delta} \times 100\%, \quad (12)$$

where R_L is the runoff change due to land use change, from time period of phase I to phase II; R_{H2} and R_{H1} are the simulated runoffs corresponding to the land uses of phase II and phase I, respectively, and both of them can be calculated with the SCS monthly model; R_C is the runoff change caused by climate change; R_B is the natural runoff of phase I; R_T is the total runoff change; R_0 is the runoff change due to other human activities except for land use change; μL, μC, and μR_0 are the percentages for the roles played by land use change, climate change, and other human activities, respectively.

RESULTS AND DISCUSSION

Time-Series Analysis

The variation trends of the climatic variables over 50 years in the study basin were analyzed with the linear regression and Mann-Kendall trend test method. Table 1 listed the detailed results obtained by the Mann-Kendall trend test method. Figure 6 demonstrated the results for mutation analysis with the Mann-Kendall trend test and the linear regression method, respectively.

Table 1: M-K trend analysis for meteorological elements in the Dongjiang basin (1958–2009)

Statistics	Meteorological elements				
	Precipitation (mm)	**Temperature (°C)**	**Evaporation (mm)**	**Humidity (g/m³)**	**Sunlight (h)**
$\overline{C}$	1852.92	21.30	1572.4	78.1	1813
C_v	0.16	0.02	0.05	0.03	0.09
C_s	0.13	0.56	0.50	−0.94	0.45
M	0.22	3.34	−2.94	−2.97	−4.18

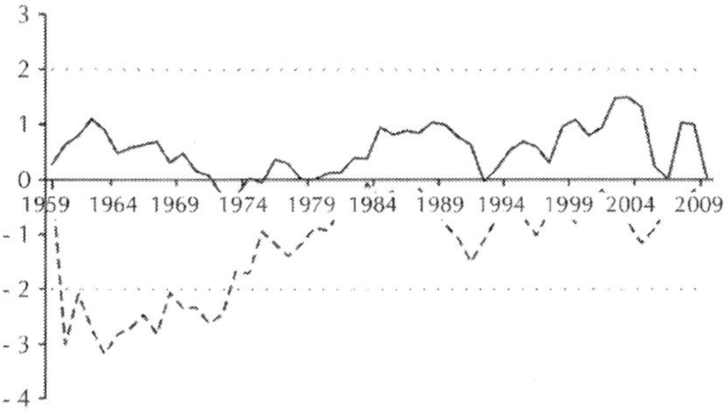

(a)

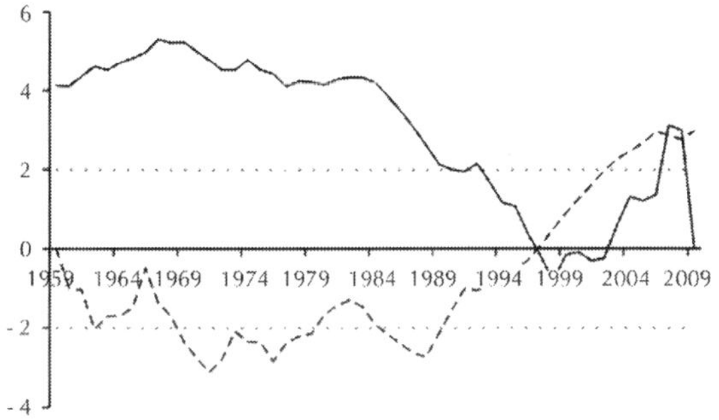

(b)

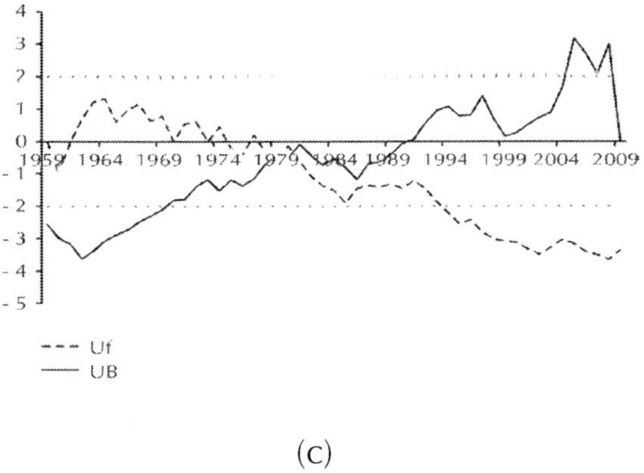

(c)

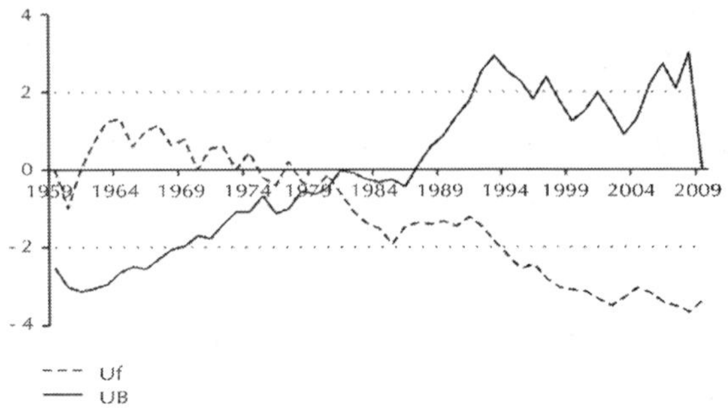

(d)

Figure 6: Mann-Kendall test trend of four meteorological elements in the Dongjing basin.

It can be seen from Table 1 that each climatic variable showed different degree of change in the Dongjiang basin during the past 50 years (1959–2008), according to its statistics by the Mann-Kendall trend test method. Precipitation showed a non-significant increasing trend (M=0.22) at 99% confidence level, which is largely due to local atmospheric circulation and topography. Temperature in the same period showed a significant increasing trend. (M=3.34) While sunlight showed a significant decreasing trend. Meanwhile, very fast urbanization caused quick increase of impervious area in the study basin in the last 50 years. These all contributed to a decreasing trend of evaporation. Temperature showed an increasing trend with a distinct mutation taking place around 1997, and sunlight and evaporation both had an abrupt change in 1982. Overall, the change of climatic variables in the Dongjiang basin during the past 50 years was significant.

It can be found from the C_v value of each climatic variable that there was a significant internal variation for precipitation but not for temperature. It can be illustrated by both the rainfall pattern and the atmospheric circulation in the study area: Dongjiang basin is located in the southern humid region in China and deeply affected by the monsoon circulation with the rainfall patterns of frontal rain and convectional rain. Therefore, the internal variation of precipitation is significant.

To clarify the relations among each climatic variable and the runoff, the correlation coefficients and relevancies were analyzed. According to the spatial differences of land cover and climatic variables, the Dongjiang basin was divided into three parts: the upper part centered with Longchuan, marked D1; the middle part centered with Heyuan, marked D2; and the whole basin centered with Boluo, marked D3. Tables 2 and 3 listed the correlation coefficients and relevancies, respectively. It can be seen from Table 2 that precipitation was positively related to runoff, and their correlation coefficient was the largest, which revealed that runoff in the Dongjiang basin depended mainly on precipitation. In contrast, temperature, sunlight, and evaporation were negatively related to runoff. Among the three parts of the basin, the correlation coefficient of precipitation and runoff in D2 was the least. The reason was that there located the biggest reservoir, the Xinfengjiang Reservoir (Figure 1) in D2. In addition, more than 20 middle and small sized reservoirs and hydraulic projects were constructed in the same period, which made a more significant change of land cover as compared to D1 and D3. Actually, runoff in D2 is mainly controlled by regulation of the reservoir, so precipitation was less correlation with runoff in D2. Table 3 showed that the relevancy between precipitation and runoff was the largest, indicating that precipitation was the major driver for runoff change in the whole study basin, which is similar to the results of correlation analysis.

Table 2: The correlation coefficient of each meteorological element to runoff in the Dongjiang basin

Basin units	Basin scope	Gauge stations	Temperature	Precipitation	Evaporation	**Sunlight**	**Humidity**
D1	Upper	Longchuan	−0.10	0.80	−0.4	−0.35	0.27
D2	Midstream	Heyuan	−0.23	0.69	−0.28	−0.39	0.35
D	The whole Basin	Boluo	−0.08	0.88	−0.54	−0.57	0.26

Table 3: The relevancies between each meteorological element and runoff in the Dongjiang basin

Basin units	Basin scope	Gauge stations	Temperature	Precipitation	Evaporation	Sunlight	Humidity
D1	Upper	Longchuan	0.4884	0.5503	0.4688	0.449	0.4702
		Rank	2	1	4	5	3
D2	Midstream	Heyuan	0.4857	0.5389	0.4659	0.4313	0.4861
		Rank	3	1	4	5	2
D	The whole basin	Boluo	0.525	0.5968	0.4883	0.4495	0.5239
		Rank	2	1	4	5	3

Land Use Change Analysis

It has been found that the urban land and farmland increased in the study area [54]. Based on the raster graphics for land use of two periods (1980 and 2000), temporal and spatial variations of land use change in the Dongjiang basin was analyzed with the ArcGIS spatial analysis technique. Figure 2 showed the maps of land use in 1980 and 2000, respectively. Table 4 depicted the detailed information of the land use change in the form of a comparison.

Table 4: Comparison of land use in the Dongjiang basin during two different period

Land use	The year 1980		The year 2000		Δ_{area} (km²)	Ratio (%)
	Area (km²)	Ratio (%)	Area (km²)	Ratio (%)		
Farmland	6083.49	22.33	7272.73	26.69	1189.24	4.36
Garden	1619.14	5.94	896.70	3.29	−722.44	−2.65
Forest	17586.88	64.56	16939.87	62.18	−647.01	−2.38
Meadow	1282.74	4.71	947.14	3.48	−335.60	−1.23
Built-up	30.02	0.12	682.25	2.5	652.23	2.38
Waters	619.99	2.28	480.12	1.79	−139.87	−0.49
Unutilized land	17.38	0.06	20.80	0.07	3.42	0.01

It can be seen from Figure 2 that the dominant land uses in the study area, in 1980, were mainly forest (widely distributed in the whole basin) and farmland (mainly distributed in the upper basin), accounting for 64.56% and 22.33%, respectively. while garden, meadow, and water took a little part with the percentage of 5.94%, 4.71%, and 2.28%, respectively. Urban land mainly distributed in the downstream basin, and the unutilized land took the least percentage in the Dongjiang basin. The area of two types of land use, urban land and farmland, increased from 1980, along with the decrease of garden (2.65%), meadow (1.23%), water (0.49%), and forest (2.38%) in 2000, while the dominant land use was still forest in 2000. In all, since 1980s, the land use in the Dongjiang basin has been characterized with a prominent increase in urban land, a little increase in farmland, and great decrease in forest area, while little change in water area and unutilized land. This land use change due to fast social and economic development had a significant impact on the hydrological response and then contributed to the variability of water resources in the region.

Runoff Simulation and Runoff Change Analysis

The Value of CN

CN (curve number) can reflect the capacity of runoff yield for the land cover with a continuous spatial distribution. The CN isocline represents the runoff yield capacities in the study basin.

Based on land use maps of two periods (1980 and 2000) and soil type data of the Dongjiang basin, CN distribution maps (see Figure 7) of 3 AMC (antecedent moisture condition) scenarios in the same periods were obtained with spatial interpolation. Figure 8 showed the CN distribution maps for AMC II of the four sub-basins in the two periods.

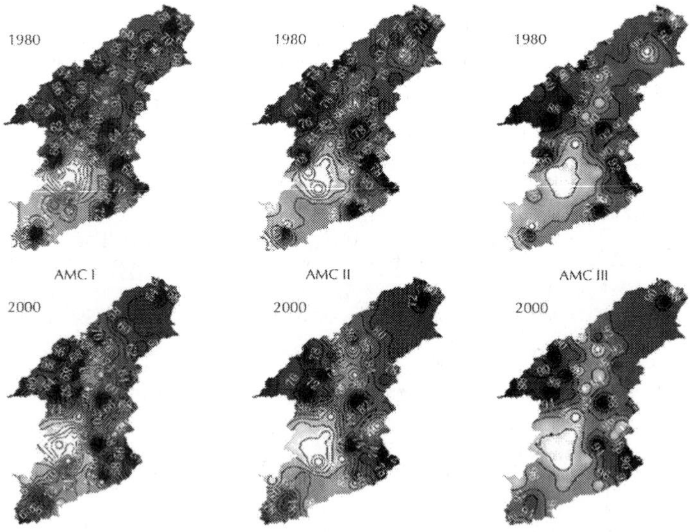

Figure 7: Comparison of CN distribution under 3 AMC scenarios in the Dongjiang basin of two periods (1980 and 2000).

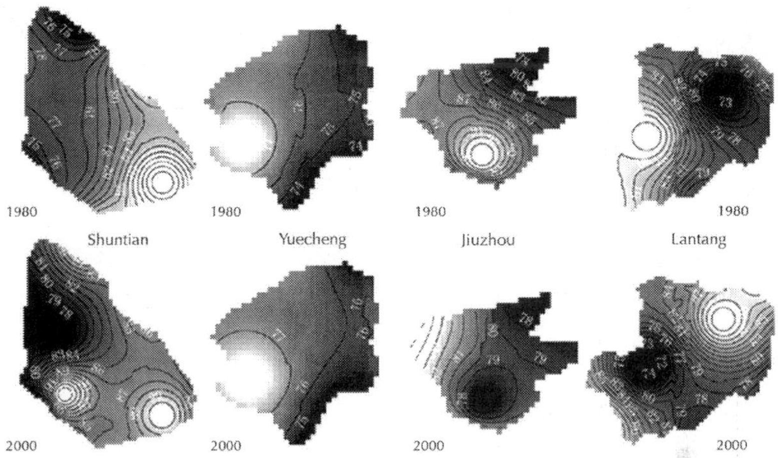

Figure 8: Comparison of the CN distribution under AMC II scenario for the four sub-basins in the Dongjiang basin of two periods (1980 and 2000).

Model Calibration and Validation

The SCS monthly model was calibrated in the period of 1970–1976 with climatic data and then validated in the period of 1977-1978 by manual calibration method with the acceptable set of parameters after CN value analysis. To measure the performance of the model, we chose Nash-Sutcliffe coefficient of efficiency (NSE) criterion [55] and relative error (RE) as objective function.

The results of simulation were listed in Table 5. The comparisons between simulated and observed runoffs of the calibration and validation periods in Shuntian sub-basin were illustrated in Figures 9 and 10, respectively. The simulated results by the SCS monthly model were comparable with the observations in the four sub-basins. Nash-Sutcliffe coefficients were above 0.8, and RE was within the pale of 0.02 for those all four sub-basins in the calibration period. It can be inferred from Figures 9 and 10 that the simulated hydrograph matched well with the observed one, and the peak flows were coincided in both the calibration and validation periods. Overall, model performance was acceptable within the study domain, and reliable to be extended to reconstruct the natural runoff.

Table 5: Calibration and validation of the SCS monthly model

Sub-basins	Runoff coefficient	Calibration			NSE/%	RE/%	Evaluation		
		Data length/ years	a	c			Data length/ years	NSE/%	RE/%
Shuntian	0.596	7	0.699	0.858	81.6	1.0	2	92.5	−9.7
Yuecheng	0.658	7	0.592	0.604	84.0	0.9	2	86.0	11.1
Lantang	0.515	7	0.601	0.832	83.0	1.6	2	94.0	−1.8
Jiuzhou	0.571	7	0.674	0.821	82.3	−0.5	2	79.3	20.9

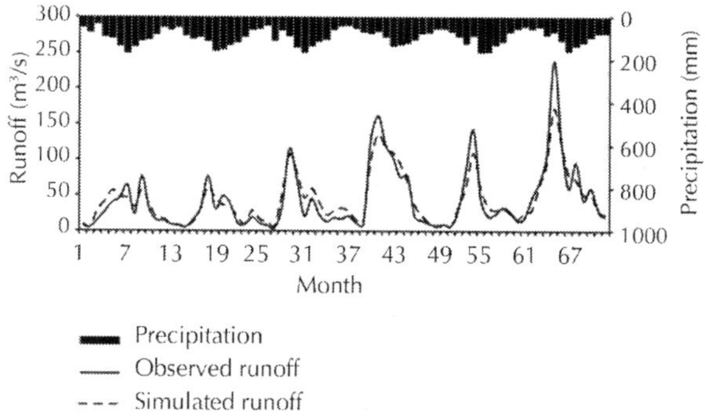

Figure 9: Comparison of simulated and observed runoffs in the calibration period (1970–1976) in the Shuntian subbasin.

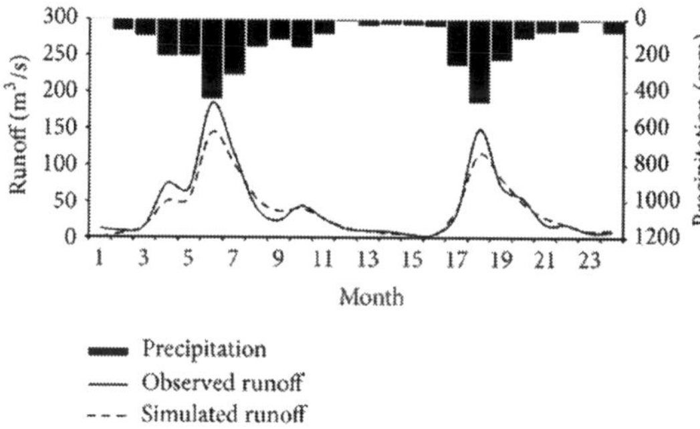

Figure 10: Comparison of simulated and observed runoffs in the validation period (1977–1978) in the Shuntian sub-basin.

Runoff Simulation in Human Activity Period

Based on the CN value of land use in the first phase (before 1980) and climatic data in the second phase (precipitation and evaporation, from 1980 to 2000) in the four sub-basins, the natural monthly runoff I and its process in the human activity period were simulated by using the SCS monthly model. And the monthly runoff II was obtained under the land use change condition (with the CN value of land use in 2000 as its input) by the same method. The difference between the simulated runoffs in the two phases was the runoff change due to land use change.

The simulated monthly runoff change due to land use change from 1980 to 2000 for each sub-basin was shown in Figure 11. It can be found that the runoff change varied from sub-basin to sub-basin, and the changes in Yuecheng sub-basin and Jiuzhou sub-basin were less than those of in Shuntian and Lantang, which was relevant to the variation scope of the CN value in the basins.

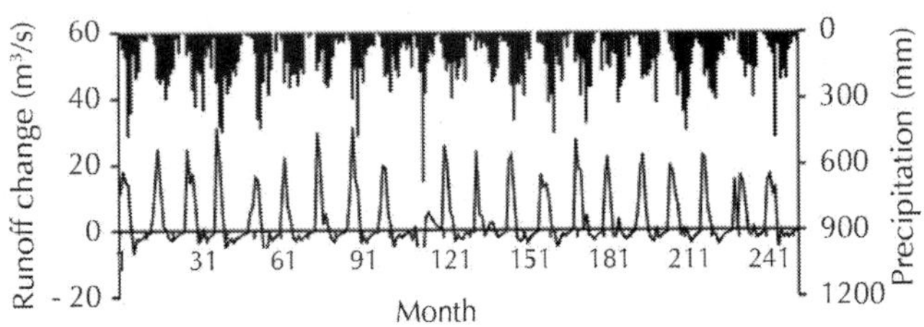

(a)

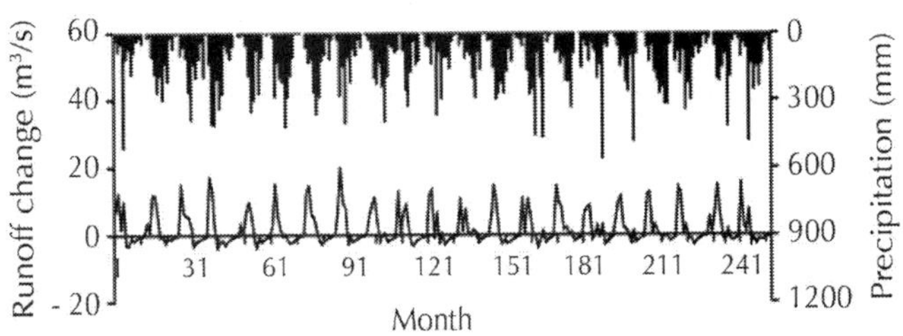

(b)

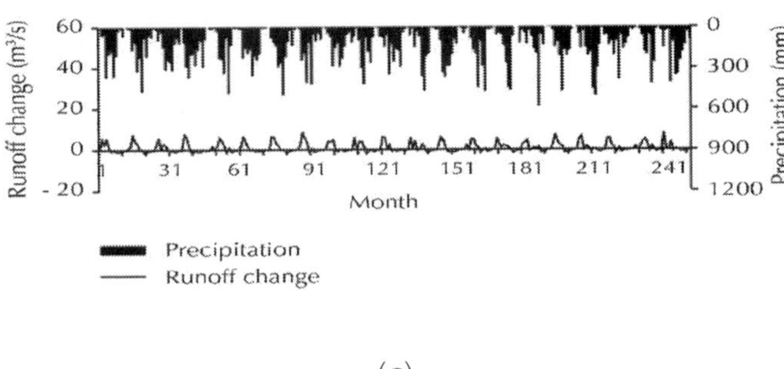

(c)

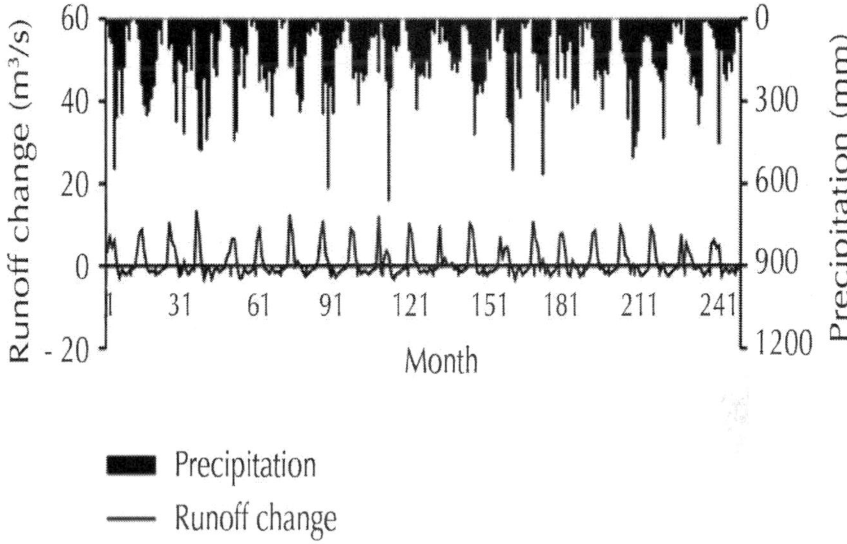

(d)

Figure 11: Processes of the simulated monthly runoffs change in the four sub-basins due to land use change from 1980 to 2000.

The Quantitative Effect due to Climate and Land Use Changes

The analysis of quantitative impacts on runoff due to human activities particularly land use and climate changes was carried out in the four sub-basins. The runoff change due to land use change was obtained from the simulated runoff by the SCS model in the two periods (human activity and natural periods). The quantitative impacts of climate change and human activities were identified by the method presented in (12), and the results were shown in Table 6.

Table 6: Monthly runoff changes of natural and human activity periods in the four sub-basins ($10^8 m^3$)

Sub-basin	Natural period R_N	R_{HR}	ΔR_T	Human activity period					
				Land use change		Climate change		Other human activities	
				ΔR_L	μL (%)	ΔR_C	μC (%)	ΔR_0	μR_0 (%)
Shuntian	1.12	1.16	0.04	0.08	24.11	−0.15	44.60	0.11	31.29
Yuecheng	0.55	0.53	−0.01	0.03	19.54	−0.08	54.74	0.04	25.72
Lantang	0.73	0.77	0.03	0.05	29.94	−0.07	40.51	0.05	29.55
Jiuzhou	0.34	0.34	−0.01	0.02	25.50	−0.05	53.22	0.02	21.27

It can be seen from Table 6 that, take Shuntian sub-basin, for instance, as compared to natural period, the runoff in human activity period increased by $4 \times 10^6 m^3$. The 24.11% increase of runoff was contributed by the land use change. On the contrary, climate change contributed the 44.6% decrease to runoff, which was superior to that of land use change. A meaningful clue has been found for explaining such a result: as two most important factors of the runoff change, precipitation showed an insignificant increasing trend, which was different from that of temperature (significant increasing trend), and the compound effects led to the reduction of the total runoff. Other factors including natural and human aspects accounted for 31.29% to the runoff change. Overall, all the driving factors can make the runoff change to a different level among which the role played by climate change took nearly the half for each sub-basin. This was in accordance with the result of identifying the quantitative effect of land use and climate change on runoff in the high flow period in the Dongjiang basin [30].

The impact of land use change on the total runoff change in Lantang sub-basin was the highest (29.94%) among the four sub-basins, while the impact ratios of climate change and other human activities were 40.51% and 29.55%, respectively. In summary, human activities contributed 59.49% to runoff change, which revealed that human activities were the major driver for runoff change.

It was obvious that climate change with increased precipitation and decreased evaporation caused the increase of runoff in the study area. Furthermore, changes of inner-annual distribution precipitation also affected the runoff change. Seven pairs of simulated runoff under nearly equal amount of precipitation were shown in Table 7. It can be found from Table 7 that the greater the percentage of precipitation in flood season, the greater the simulated runoff in Shuntian sub-basin, except for 1985 and 1987. It was also implied that precipitation was the major driver for runoff change in Shuntian sub-basin

Table 7: Simulated runoff yielded by similar amount of precipitations in the three sub-basins

Subbasins	Year	Precipitation (mm)			Simulated runoff ($10^8 m^3$)
		Annual	Flood season	Percentage of flood season (%)	
Shuntian	2004	1274	1108.10	86.57	7.56
	1999	1279	1107.50	87.00	8.41
	2002	1421	1079.61	76.00	8.12
	2003	1425	1256.16	88.17	9.80
	1985*	1639	1217.40	74.27	11.83
	1987*	1643	1185.57	72.14	13.15
	1982	1811	1396.43	77.11	13.04
	1984	1813	1592.17	87.83	14.70
Lantang	1986	1628	1259.60	77.39	7.96
	1998	1631	1148.10	70.39	8.77
Jiuzhou	1981	1754	1427.00	81.35	4.00
	1982	1752	1398.10	79.79	3.48
	1984	1577	1498.50	95.02	3.87
	1989	1575	1316.70	83.61	3.51

Annual precipitation for Yuecheng subbasin varies greatly all the years as compared with other sub-basins, and no similar amount of precipitations can be found in different years. The impact of inner-annual distribution of precipitation on the variation of runoff in Yuecheng subbasin was neglected.

To analyze the impact of land use change, we divided annual precipitation into four classes as under 1500 mm, 1500–1800 mm, 1800–2000 mm, and above 2000 mm in the four sub-basins. The total runoff change due to land use change in each class was shown in Figure 12. It can be found that under larger precipitation, the runoff (to be flattened) of flood season was less as compared with that of drought season and the whole year in the four sub-basins. This was attributed to land cover conversion due to construction of reservoirs, which stored much water in flood season for flood prevention and water use. In addition, runoff yield changed to be much quick and more sensitive to precipitation, along with the land cover conversion from forest to urban areas or other vegetation types. The impact of land use change on runoff was more distinctive when the precipitation was larger.

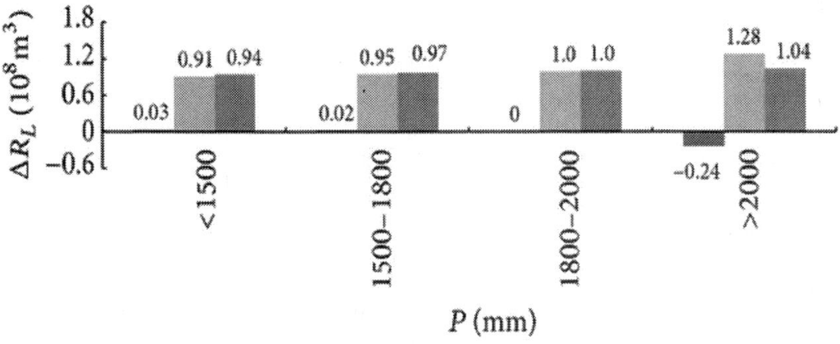
ΔR_L (10^8m^3)
1.8
1.2
0.6
0
−0.6
0.03 0.91 0.94
0.02 0.95 0.97
0 1.0 1.0
−0.24 1.28 1.04
<1500
1500–1800
1800–2000
>2000
P (mm)

(a)

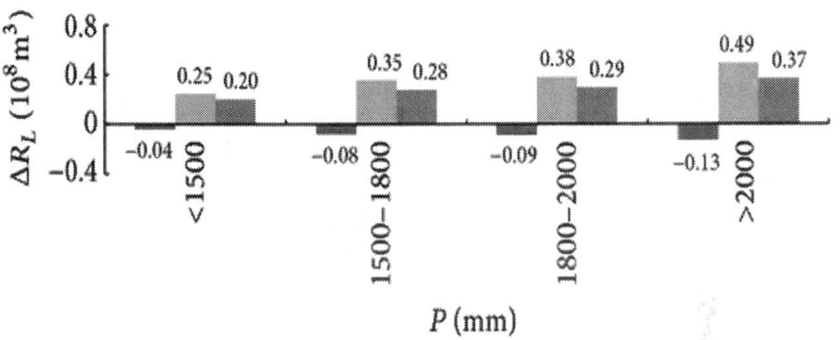
ΔR_L (10^8m^3)
0.8
0.4
0
−0.4
−0.04 0.25 0.20
−0.08 0.35 0.28
−0.09 0.38 0.29
−0.13 0.49 0.37
<1500
1500–1800
1800–2000
>2000
P (mm)

(b)

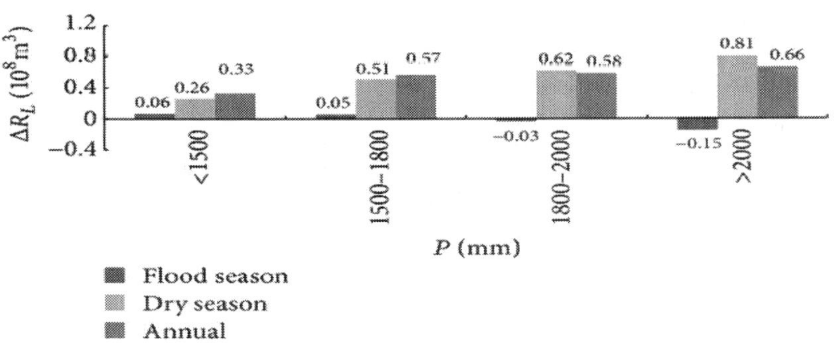
ΔR_L (10^8m^3)
1.2
0.8
0.4
0
−0.4
0.06 0.26 0.33
0.05 0.51 0.57
−0.03 0.62 0.58
−0.15 0.81 0.66
<1500
1500–1800
1800–2000
>2000
P (mm)
Flood season
Dry season
Annual

(c)

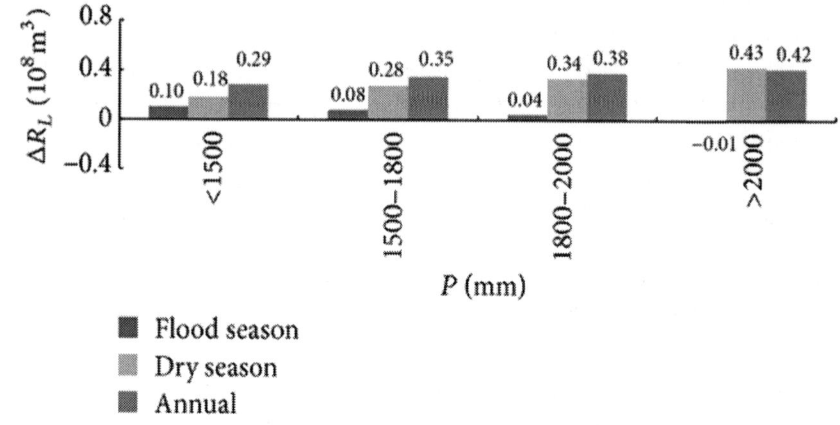

(d)

Figure 12: Average annual runoff change due to land use change under different precipitation classes in the four sub-basins.

Finally, it can be concluded that the runoff change was affected by many factors, and the contribution ratio of each factor was different. Climate change and human activities (especially the land use change) were the two dominant drivers which contributed about 40%~50% each to the runoff change. Particularly the land use change, whose impact on the runoff change in flood season under large precipitation, should not be neglected. The role played by climate change and human activities including land use change should be considered, respectively, in the analysis of variability and availability of water resources, and then reasonable measures and policies should be taken.

CONCLUSIONS

Based on land use maps of two time periods in the Dongjiang basin, this study identified the quantitative effects of land use and climate change on the runoff, which revealed some interesting results with the application of the SCS monthly model.

1. The climate experienced significant changes in the basin over the past 50 years, and the changes were detected using the Mann-Kendall test method with the results of nonsignificant increase in

precipitation, significant increase in temperature while decrease in evaporation and sunlight.

2. Land use has experienced a significant change since the 1980s in the Dongjiang basin with the characteristic of spatial distribution for the CN value under three AMC scenarios in two periods. The average value of CN increased and the CN value varied in each sub-basin due to different land use patterns. In particular, expansion of urban area in the south part of the study area (downstream) and deforestation in the whole area provided contributory factors that affected hydrological processes and subsequently increased runoff.
3. The SCS monthly model performed well in the four sub-basins of the Dongjiang basin, and the results showed that the runoff change in each sub-basin during two time periods was different. The runoff change in Yuecheng sub-basin was less than that of Shuntian sub-basin and Lantang sub-basin, which was in good correlation to the variation scope of the CN value. The more land use changed, the greater the CN value changed, which resulted in the greater variation of the runoff under the same climatic condition.
4. The separated quantitative effect of land use and climate change on the runoff showed that climate change and human activities (including land use) contributed half and half, respectively, to runoff change. While land use change independently contributed 20%~30% to the total runoff change. Moreover, the effect of land use on runoff change was in different level under different amounts of precipitation. The effect of climate change on runoff change had a different inner-annual distribution even under a same amount of annual precipitation.

Further research is required to acquire the regional future climate scenarios coupled with the hydrological model of a basin scale under GCMs (general circulation models) with the downscaling technique, so as to further quantify the relations between runoff and climatic variables. In addition, the space-time distribution of floods and droughts resulted from the runoff change should also be studied to provide scientific framework for basin-scale water resources management.

ACKNOWLEDGMENTS

The research is financially supported by the National Natural Science Foundation of China (Grant no. 51210013), the Public Welfare Project of Ministry of Water Resources (Grant no. 201301071-02, 201301002-02), the project from Guangdong Science and Technology Department (Grant no. 2010B050300010), the project from Guangdong Water Resources Department (Grant no. 2009-39), Key Project of Chinese National Programs for Fundamental Research and Development (973 program) (Grant no. 2013CB036406, 2010CB951102), and the Pearl-River-New-Star of Science and Technology supported by Guangzhou City (Grant no. 2011J2200051).

REFERENCES

1. A. Bronstert, D. Niehoff, and G. Brger, "Effects of climate and land-use change on storm runoff generation: present knowledge and modelling capabilities," Hydrological Processes, vol. 16, no. 2, pp. 509–529, 2002.
2. D. Legesse, C. Vallet-Coulomb, and F. Gasse, "Hydrological response of a catchment to climate and land use changes in Tropical Africa: case study south central Ethiopia," Journal of Hydrology, vol. 275, no. 1-2, pp. 67–85, 2003.
3. B. Zimmermann, H. Elsenbeer, and J. M. De Moraes, "The influence of land-use changes on soil hydraulic properties: implications for runoff generation," Forest Ecology and Management, vol. 222, no. 1–3, pp. 29–38, 2006.
4. D. Mao and K. A. Cherkauer, "Impacts of land-use change on hydrologic responses in the Great Lakes region," Journal of Hydrology, vol. 374, no. 1-2, pp. 71–82, 2009.
5. N. A. Adnan and P. M. Atkinson, "Exploring the impact of climate and land use changes on streamflow trends in a monsoon catchment," International Journal of Climatology, vol. 31, no. 6, pp. 815–831, 2011.
6. T. Jiang, Y. D. Chen, C.-Y. Xu, X. Chen, X. Chen, and V. P. Singh, "Comparison of hydrological impacts of climate change simulated

by six hydrological models in the Dongjiang Basin, South China," Journal of Hydrology, vol. 336, no. 3-4, pp. 316–333, 2007.

7. G. B. Fu, J. J. Yu, X. B. Yu et al., "Temporal variation of extreme rainfall events in China, 1961–2009,"Journal of Hydrology, vol. 487, pp. 48–59, 2013.
8. G. A. Meehl, F. Zwiers, J. Evans, T. Knutson, L. Mearns, and P. Whetton, "Trends in extreme weather and climate events: issues related to modeling extremes in projections of future climate change," Bulletin of the American Meteorological Society, vol. 81, no. 3, pp. 427–436, 2000.
9. S. Wang, S. Kang, L. Zhang, and F. Li, "Modelling hydrological response to different land-use and climate change scenarios in the Zamu River basin of northwest China," Hydrological Processes, vol. 22, no. 14, pp. 2502–2510, 2008.
10. M. Verbunt, M. Groot Zwaaftink, and J. Gurtz, "The hydrologic impact of land cover changes and hydropower stations in the Alpine Rhine basin," Ecological Modelling, vol. 187, no. 1, pp. 71–84, 2005.
11. M. D. Tomer and K. E. Schilling, "A simple approach to distinguish land-use and climate-change effects on watershed hydrology," Journal of Hydrology, vol. 376, no. 1-2, pp. 24–33, 2009.
12. E.-S. Chung, K. Park, and K. S. Lee, "The relative impacts of climate change and urbanization on the hydrological response of a Korean urban watershed," Hydrological Processes, vol. 25, no. 4, pp. 544–560, 2011.
13. S. Piao, P. Ciais, Y. Huang et al., "The impacts of climate change on water resources and agriculture in China," Nature, vol. 467, no. 7311, pp. 43–51, 2010.
14. P. Zhai, X. Zhang, H. Wan, and X. Pan, "Trends in total precipitation and frequency of daily precipitation extremes over China," Journal of Climate, vol. 18, no. 7, pp. 1096–1108, 2005.
15. Y. Q. Chen, Q. Zhang, X. H. Chen, and P. Wang, "Multi-scale variability of runoff changes in the Pearl River basin, China," Quaternary International, vol. 226, pp. 44–53, 2010.
16. C. O. Wilson and Q. Weng, "Simulating the impacts of future land use and climate changes on surface water quality in the Des Plaines River watershed, Chicago Metropolitan Statistical Area,

Illinois," Science of the Total Environment, vol. 409, no. 20, pp. 4387–4405, 2011.

17. L. Cuo, T. K. Beyene, N. Voisin et al., "Effects of mid-twenty-first century climate and land cover change on the hydrology of the Puget Sound basin, Washington," Hydrological Processes, vol. 25, no. 11, pp. 1729–1753, 2011.
18. J. I. López-Moreno, S. M. Vicente-Serrano, E. Moran-Tejeda, J. Zabalza, J. Lorenzo-Lacruz, and J. M. García-Ruiz, "Impact of climate evolution and land use changes on water yield in the ebro basin,"Hydrology and Earth System Sciences, vol. 15, no. 1, pp. 311–322, 2011.
19. L. M. Mango, A. M. Melesse, M. E. McClain, D. Gann, and S. G. Setegn, "Land use and climate change impacts on the hydrology of the upper Mara River Basin, Kenya: results of a modeling study to support better resource management," Hydrology and Earth System Sciences, vol. 15, no. 7, pp. 2245–2258, 2011.
20. J. C. I. Dooge, M. Bruen, and B. Parmentier, "A simple model for estimating the sensitivity of runoff to long-term changes in precipitation without a change in vegetation," Advances in Water Resources, vol. 23, no. 2, pp. 153–163, 1999.
21. R. D. Koster and M. J. Suarez, "A simple framework for examining the interannual variability of land surface moisture fluxes," Journal of Climate, vol. 12, no. 7, pp. 1911–1917, 1999.
22. P. C. D. Milly and K. A. Dunne, "Macro-scale water fluxes: water and energy supply control of their inter-annual variability," Water Resources Research, vol. 38, p. 1206, 2002.
23. X. C. Ye, Q. Zhang, and J. Liu, "Impact of climate change and human activities on runoff of Poyang Lake Catchment," Journal of Glaciology and Geocryology, vol. 31, no. 5, pp. 835–842, 2009 (Chinese).
24. X. Ma, J. Xu, and M. van Noordwijk, "Sensitivity of streamflow from a Himalayan catchment to plausible changes in land cover and climate," Hydrological Processes, vol. 24, no. 11, pp. 1379–1390, 2010.
25. X. Hao, Y. Chen, C. Xu, and W. Li, "Impacts of climate change and human activities on the surface runoff in the Tarim River Basin over the last fifty years," Water Resources Management, vol. 22, no. 9, pp. 1159–1171, 2008.

26. J. Wang, Y. Hong, J. Gourley, P. Adhikari, L. Li, and F. Su, "Quantitative assessment of climate change and human impacts on long-term hydrologic response: a case study in a sub-basin of the Yellow River, China," International Journal of Climatology, vol. 30, no. 14, pp. 2130–2137, 2010.
27. Z. Li, W.-Z. Liu, X.-C. Zhang and F.-L. Zheng, "Impacts of land use change and climate variability on hydrology in an agricultural catchment on the Loess Plateau of China," Journal of Hydrology, vol. 377, no. 1-2, pp. 35–42, 2009.
28. G. Wang, J. Xia, and J. Che, "Quantification of effects of climate variations and human activities on runoff by a monthly water balance model: a case study of the Chaobai River basin in northern China,"Water Resources Research, vol. 45, no. 7, Article ID W00A11, 2009.
29. J. Fan, F. Tian, Y. Yang, S. Han, and G. Qiu, "Quantifying the magnitude of the impact of climate change and human activity on runoff decline in Mian River Basin, China," Water Science and Technology, vol. 62, no. 4, pp. 783–791, 2010.
30. D. Liu, X. Chen, Y. Lian, and Z. Lou, "Impacts of climate change and human activities on surface runoff in the Dongjiang River basin of China," Hydrological Processes, vol. 24, no. 11, pp. 1487–1495, 2010.
31. K. Lin, S. Guo, W. Zhang, and P. Liu, "A new baseflow separation method based on analytical solutions of the Horton infiltration capacity curve," Hydrological Processes, vol. 21, no. 13, pp. 1719–1736, 2007.
32. K. Lin, Q. Zhang, and X. Chen, "An evaluation of impacts of DEM resolution and parameter correlation on TOPMODEL modeling uncertainty," Journal of Hydrology, vol. 394, no. 3-4, pp. 370–383, 2010.
33. Y. Hundecha and A. Bárdossy, "Modeling of the effect of land use changes on the runoff generation of a river basin through parameter regionalization of a watershed model," Journal of Hydrology, vol. 292, no. 1–4, pp. 281–295, 2004.
34. K. Y. Li, M. T. Coe, N. Ramankutty, and R. D. Jong, "Modeling the hydrological impact of land-use change in West Africa," Journal of Hydrology, vol. 337, no. 3-4, pp. 258–268, 2007

35. G. Parkin, G. O'Donnell, J. Ewen, J. C. Bathurst, P. E. O'Connell, and J. Lavabre, "Validation of catchment models for predicting land-use and climate change impacts. 2. Case study for a Mediterranean catchment," Journal of Hydrology, vol. 175, no. 1–4, pp. 595–613, 1996.

36. J. K. Lørup, J. C. Refsgaard, and D. Mazvimavi, "Assessing the effect of land use change on catchment runoff by combined use of statistical tests and hydrological modelling: case studies from Zimbabwe,"Journal of Hydrology, vol. 205, no. 3-4, pp. 147–163, 1998.

37. D. Niehoff, U. Fritsch, and A. Bronstert, "Land-use impacts on storm-runoff generation: scenarios of land-use change and simulation of hydrological response in a meso-scale catchment in SW-Germany,"Journal of Hydrology, vol. 267, no. 1-2, pp. 80–93, 2002.

38. B. Troy, C. Sarron, J. M. Fritsch, and D. Rollin, "Assessment of the impacts of land use changes on the hydrological regime of a small rural catchment in South Africa," Physics and Chemistry of the Earth, vol. 32, no. 15–18, pp. 984–994, 2007.

39. H. Chen, S. Guo, C.-Y. Xu, and V. P. Singh, "Historical temporal trends of hydro-climatic variables and runoff response to climate variability and their relevance in water resource management in the Hanjiang basin," Journal of Hydrology, vol. 344, no. 3-4, pp. 171–184, 2007.

40. D. H. Burn, "Climatic influences on streamflow timing in the headwaters of the Mackenzie River Basin,"Journal of Hydrology, vol. 352, no. 1-2, pp. 225–238, 2008.

41. H. B. Mann, "Nonparametric tests against trend," Econometrica, vol. 13, no. 7, pp. 245–259, 1945.

42. M. G. Kendall, Rank Correlation Methods, Griffin, London, UK, 1975.

43. J. M. Mitchell, B. Dzerdzeevskii, H. Flohn et al., "Climate change," WMO Technical Note 79, World Meteorological Organization, 1966.

44. E. Kahya and S. Kalayci, "Trend analysis of streamflow in Turkey," Journal of Hydrology, vol. 289, no. 1–4, pp. 128–144, 2004.

by impacting the evapotranspiration, soil infiltration capacity, and surface and subsurface flow regimes [8, 10, 11], while climate change can alter the flow routing time and peak flows [10, 12]. It is crucial to the long-term water resource planning and management to better understand the potential impacts of climate and LULC changes on the runoff and streamflow in basins [9, 13]. In particular, effective water resource management under changing conditions requires reliable information about flows and modes that can be used to simulate flow regimes under different scenarios of changing land use and climate [6].

Separation of impacts of climate and LULC changes on the hydrological cycle is of great importance to improving the land use planning and water resource management [10, 14], especially in arid and semiarid regions where the climate change may significantly affect the hydrological cycle [15]. The effects of LULC and climate changes on the streamflow are more evident in the arid and semiarid regions. One typical example is the Heihe River Basin in Northwest China, which is characterized by limited water resources and special hydroclimatic and physiographic conditions [16]. Understanding the hydrological responses to potential climate change is very important for developing sustainable water resource management strategies in this region. However, the impacts of urbanization and deforestation on the hydrological cycle in arid and semiarid regions have been rarely documented [17–19]. Overall, there is still very limited understanding of the separate as well as combined impacts of LULC and climate changes on regional water and energy cycles, and therefore more in-depth research is needed, especially in the arid and semiarid regions [17].

There are many studies about the hydrological impacts of land use change or climate change at the basin scale, and most of them were conducted with a hydrological model based on a series of land use data extracted from satellite images [9, 20, 21]. For example, the impacts of land use change scenarios in the Wutu watershed, North Taiwan, were assessed using the conversion of a land use model (CLUE-s) and a generalized watershed loading functions model [16]. A Soil Water Assessment Tool (SWAT) and multiple General Circulation Models (GCMs) were used to investigate the relationship between climatic and hydrological changes in the Upper Mississippi River [22]. Most of these studies assume no change in LULC [4, 23], but the impacts of climate change on hydrology vary among regions and should be investigated with regional climate change scenarios [4, 24]. Besides, the hydrological impacts of LULC change also vary with the climatic conditions [9]. For example, water balance variables might add or subtract the impacts of climate change under varying land cover conditions. In particular, the regional LULC change can offset or magnify the changes in global average temperature and can significantly alter the impacts associated with global warming [17, 25]. In addition, some studies about the combined effects of climate and LULC changes on streamflow showed that climate change was generally more significant than LULC change in determining the basin hydrological response [26–28]. For example, climate dominates the changing streamflow in the Xinjiang River Basin of Poyang Lake, China [26]. However, the hydrological cycle in a basin is a complex process influenced by climate and the physical properties of the catchment and human activities together [4, 5]. The complexity of these factors complicates the separation between effects of land use and climatic variability on streamflow [27, 29]. Therefore, it is still a challenge to distinguish the effects of LULC change from that of concurrent climate variability [14].

This study aims to separate the impacts of climate and LULC changes on the hydrological cycle in the Heihe River Basin under future scenarios to provide some useful reference information that can be used to improve the water resource management and guarantee the sustainable development. The climatic data predicted by General Circulation Models (GCMs) under RCP 4.5 scenario were used to represent the climate change scenarios for 2006–2030, and the land use data simulated with the Dynamic Land Use System (DLS) model were used to represent the land use change scenarios. The future hydrological cycle was simulated with the SWAT based on the scenario data of climate change and land use change, the hydrological impacts of which were analyzed by comparing the simulation results under different scenarios. The results of this study can provide valuable information for guiding future water resource management in the Heihe River Basin as well as other arid and semiarid regions in China.

METHODS

Study Area

The Heihe River Basin is the second largest inland river basin in China, which lies between 37°43′–42°41′N and 97°23′–102°72′E with a total area of 127,96 thousand km^2. This basin expands across Qinghai Province, Gansu Province, and Inner Mongolia Autonomous Region in Northwest China (Figure 1). With a total length of 821 km, the Heihe River is divided into the upper, middle, and lower reaches, where the natural and socioeconomic characteristics differ significantly. For example, the average annual precipitation is between 200 and 500 mm, less than 200 mm, and less than 50 mm in these reaches, respectively, while the annual evaporation ranges from 700 mm in the upper reach to more than 3000 mm in the lower reach [30]. Besides, the annual average temperature is 9.4°C over the last 30 years, and this basin enjoys a dry continental climate. The altitude ranges from 869 to 5542 m, with an average of 1778 m. The main land cover types are desert (57.15% of total basin area), mountains (33.16%), and oasis (8.19%) [31]. The ecosystems from the upper reach to the lower reach are linked by the hydrological cycle, but the hydrological cycle has significantly changed due to the land use change and climate change in the past decades. For example, about 65% of the irrigation water in the middle reach was extracted from the river runoff, which greatly influences the hydrological cycle of the whole basin. Therefore, a detailed and integrated simulation analysis of the water resources is critical and urgent for better water resource management in the Heihe River Basin.

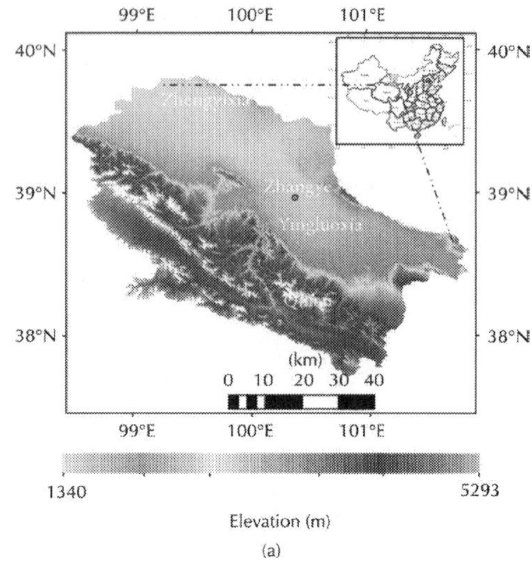
99°E
100°E
101°E
40°N
39°N
38°N
Zhangye
Yingluoxia
(km)
0 10 20 30 40
1340
5293
Elevation (m)
(a)

(a)

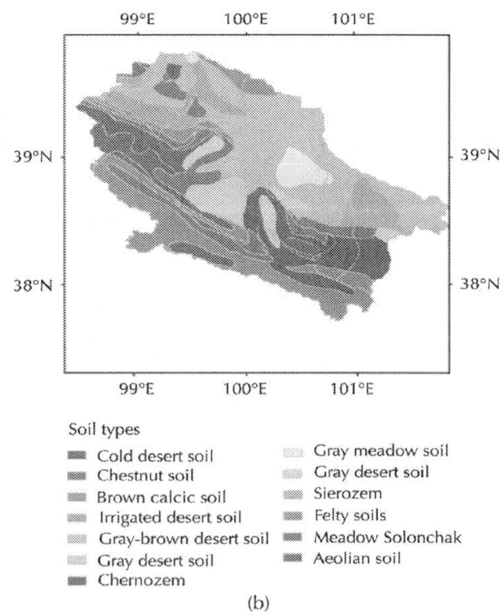
99°E
100°E
101°E
39°N
38°N
Soil types
Cold desert soil
Chestnut soil
Brown calcic soil
Irrigated desert soil
Gray-brown desert soil
Gray desert soil
Chernozem
Gray meadow soil
Gray desert soil
Sierozem
Felty soils
Meadow Solonchak
Aeolian soil
(b)

(b)

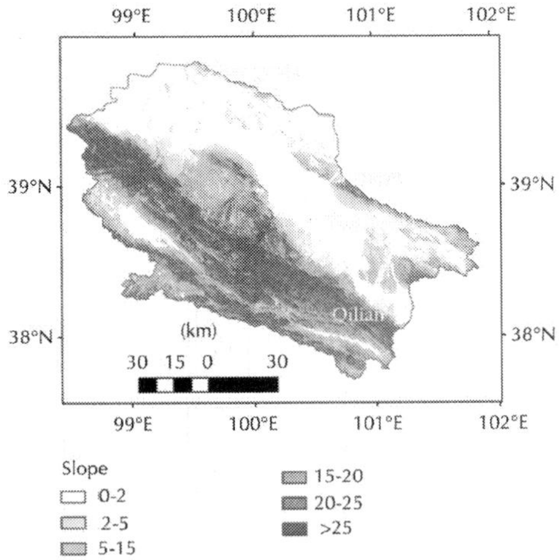

(c)

(c)

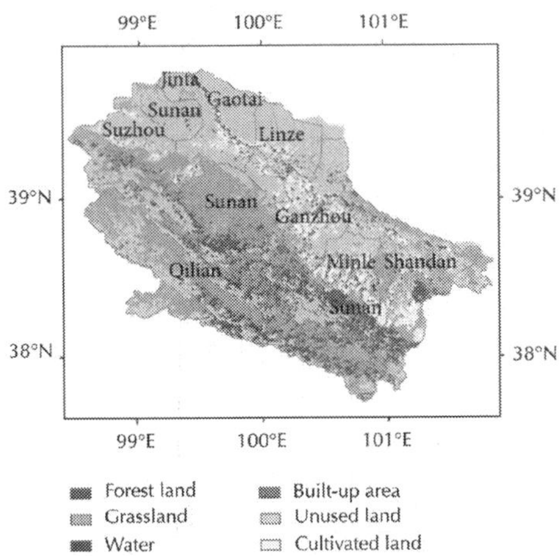

(d)

(d)

Figure 1: The upper and middle Heihe River Basin: (a) digital elevation model, (b) soil type, (c) slope, and (d) land use in 2010.

Data for Model Simulation

The spatial data (i.e., topography, soil and land use), historical climate data, and hydrological data for the watershed were first prepared for the SWAT model and DLS model. The topography was represented with the 90 m resolution digital elevation model (DEM) of Shuttle Radar Topography Mission (SRTM) (http://srtm.csi.cgiar.org/) [32]. The soil data, including texture, depth, and drainage attributes, were from the Harmonized World Soil Database (HWSD) supplied by the Environmental and Ecological Science Data Center for West China (WestDC) (http://westdc.westgis.ac.cn/). The historical land use data including 25 land use types, which were derived from Landsat TM/ETM images, were provided by Data Center of Chinese Academy of Sciences (CAS) [33]. In particular, the glacier data were obtained from WestDC (http://westdc.westgis.ac.cn/), and land use properties were directly obtained from the SWAT model database (Table 1). The historical hydrological data for SWAT model calibration and validation include the river flow data from four hydrological stations. The hydrological observation data, including annual data in Heihe River Basin during 1980–2010, were obtained from the hydrological yearbook provided by WestDC. The river flow data used for the model calibration and validation were provided by the Data Center of CAS. The historical daily weather data during 1980–2010 were collected from China Meteorological Administration (CMA), including daily precipitation, maximum and minimum temperatures, solar radiation, humidity, wind speed, and wind direction from 13 weather stations in or near the Heihe River Basin.

Table 1: Data used and sources

Data type	Data sources	Scale	Description
DEM	Shuttle Radar Topography Mission (SRTM)	90 m	Elevation

Land use	Data Center of Chinese Academy of Science	1:100000	The classification system contains 25 categories
Soil	WestDC (http://westdc.westgis.ac.cn/)	1:1000000	Some parameters were calculated using a (Soil-Plant-Atmosphere-Water) Field and Pond Hydrology model
Weather	China Meteorological Administration	Daily	5 weather stations
Hydrological observation	Hydrological yearbook	Daily	4 stations
River flow	Data Center of Chinese Academy of Science	1:250000	River network diversion
Glacier	WestDC (http://westdc.westgis.ac.cn/)	1:100000	The attributes include width, length, and depth

Simulation of Land Use Changes with the DLS Model

The DLS model is a set of applications used to simulate the changing process of land use system and is an effective tool for simulation of spatial-temporal land use changes to assist land management [34]. The DLS model consists of three modules: a spatial regression module that identifies the relationships between land uses and the influencing factors; a scenario analysis module of land use changes that determines the land demands at the regional level; a spatial disaggregation module that allocates land use changes from a regional level to the disaggregated grid cells [35, 36].

The DLS model simulates the spatial-temporal land use changes with three processes: scenario analysis of land use change, spatial regression analysis, and spatial allocation of land use changes. The first process was carried out with the scenario analysis module, which provides the data of total land demands at the annual scale during a given period. By including the scenario analyses of land use changes, a set of spatially explicit simulation results of land use change can be exported by the DLS model [37]. The total land use demands can be set by several approaches, such as trend analysis methods (e.g., linear interpolations or more sophisticated econometric models) and economic models. In this study, the total land demands during the simulation period were first determined using trend analysis methods and then were used to establish a scenario of land use change during 2006–2030.

In the spatial regression module, the relationships between land uses and influencing factors were analyzed via stepwise logistic regression analysis of past land use changes and their drivers [38]. For each grid cell, the total probability for each land use type is calculated on basis of the multinomial logistic regression at the pixel scale as follows:

$$P_{i,k} = \begin{cases} \dfrac{1}{1 + \sum_{k \neq i} e^{x_k \beta_k}} & i = k \\ \dfrac{e^{x_k \beta_k}}{1 + \sum_{k \neq i} e^{x_k \beta_k}} & i \neq k, \end{cases} \tag{1}$$

where $P_{i,k}$ is the probability of conversion from land type i to k in the cells under given driving factors; $X_1, X_2, \ldots, X_n$ represent the driving factors of climate, landform, location, population, economic growth, policy, and other categories; $\beta_1, \beta_2 \cdots \beta n$ are the regression analysis coefficients of driving factors for further estimation.

In this study, all the data of land use and the influencing factors were prepared at the annual scale. The land use data in 2000 and 2005 were used in the logistic regression. The assumed driving factors were categorized into five groups: climate, geophysics, transportation, location, and socioeconomics. The data of these factors in the corresponding years were also prepared (Table 2).

Table 2: Results of logistical regression for the six land use types with 13 driving factors

	Cultivated land	Forest land	Grassland	Water	Built-up area	Unused land
Slope	−0.000049 (−27.52)***	0.000121 (−63.65)***	−7.42E − 05 (−18.89)***	−7.76E − 06 (−7.24)***	−1.1E − 06 (−2.07)*	0.0000105 (−2.64)**
Elevation	−0.000136 (−53.82)***	−3.02E − 05 (−11.13)***	0.000116 (−20.8)***	−0.0000149 (−9.76)***	−5.9E − 06 (−7.85)***	0.0000711 (−12.56)***
Rain	−0.000045 (−16.09)***	0.0000228 (−7.61)***	0.0000419 (−6.79)***	0.0000136 (−8.05)***	1.4E − 06 −1.7	−0.0000347 (−5.54)***
Radiation	−0.000407 (−24.27)***	0.000243 (−13.55)***	−0.000923 (−24.98)***	−8.64E − 06 (−0.86)	1.16E − 05 (−2.34)*	0.00108 (−28.91)***
>0°C accumulated temperature	−0.0000106 (−24.03)***	0.0000127 (−26.78)***	0.000026 (−26.59)***	−3.88E − 06 (−14.55)***	−7.2E − 07 (−5.52)***	−0.0000234 (−23.67)***
Soil organic matter	0.0685 (−9.38)***	0.232 (−29.64)***	−0.186 (−11.58)***	0.0152 (−3.46)***	−0.00655 (−3.03)**	−0.123 (−7.52)***
pH value	0.00769 (−7.15)***	−0.0131 (−11.32)***	−0.00541 (−2.28)*	0.00171 (−2.64)**	0.000719 (−2.26)*	0.00834 (−3.47)***
Population density	0.000247 (−32.18)***	−7.82E − 05 (−9.52)***	−3.16E − 05 (−1.87)	0.0000206 (−4.46)***	−3.2E − 06 (−1.40)	−0.000154 (−9.00)***
Per capita GDP	0.000548 (−76.02)***	−3.08E − 05 (−3.99)***	−0.000185 (−11.62)***	−0.00002 (−4.61)***	0.000411 (−192.54)***	−0.000723 (−44.87)***
Distance to railway	0.000193 (−13.55)***	0.000229 (−14.96)***	0.000979 (−31.16)***	0.00000653 −0.76	1.72E − 05 (−4.07)***	−0.00142 (−44.71)***
Distance to road	−0.000221 (−18.45)***	−0.000335 (−26.06)***	0.00111 (−42.08)***	−0.000066 (−9.13)***	−1.4E − 05 (−3.79)***	−0.000477 (−17.79)***
Distance to water source	−0.000338 (−14.46)***	0.0000239 −0.95	0.000926 (−17.96)***	−0.000346 (−24.55)***	−5.2E − 06 (−0.75)	−0.000261 (−4.99)***
Distance to city	−0.000125 (−8.11)***	−0.000108 (−6.58)***	−0.000567 (−16.76)***	0.0000996 (−10.78)***	−2E − 05 (−4.31)***	0.00072 (−20.98)***

Note: t statistics in parentheses; *significant at 10%; **significant at 5%; ***significant at 1%.

The spatial disaggregation module is used to spatially and explicitly convert the land demands into land use changes at various locations of the study area. The spatial disaggregation is carried out in an iterative procedure based on the probability maps, conversion rules, historical land use maps, and land demands under the scenarios. The probability maps of each land use type were prepared with the logistic regression results. Besides, the rules of land use conversion were set for each land use type, whose value ranges from 0 to 1. A smaller value means one land use type is more likely to be converted to another type, and vice versa. The development-restricted areas in the study area were also specified.

Downscaling of GCM Climate Data

GCMs are arguably the best available tools for modeling future climate. Yet GCMs provide information at a resolution that is too coarse to be directly used in hydrological modeling [39]. Therefore downscaling is required to transform the low resolution GCM outputs to the high resolution climate features needed for hydrological simulation. The downscaling procedure is as follows. First, the average annual precipitation and temperature of 30 years (1980–2010) were calculated, which were adopted as a baseline for selecting the GCMs. We considered future climate change scenarios for the basin (Figure 2) by using the spatially distributed outputs from 10 GCMs under RCP 4.5 scenario. The climate projections of Max Planck Institute (MPI) were downscaled to the 3 km × 3 km grid in the study area, and bias was corrected and climate change scenarios were developed by MPI for Meteorology. The annual mean values of 10 GCM spatial data from 2006 to 2030 were calculated according to the basin perimeter, and one out of ten GCMs was also selected based on the historical trends and annual averages of temperature and precipitation. The MPI model was finally chosen through comparison. The result of MPI model originates in MPI, and the spatial resolution is 1.865° (LAT) × 1.875° (LON). Then the parameters in MPI GCM were transformed into the forcing data of a regional climate model in weather research and forecasting (WRF) simulation, and thereafter dynamical downscaling simulation was performed at the spatial resolution of 3 km for the period of 2006–2030. We considered the impacts of land use changes on regional climate, and the land cover data before WRF simulation were dynamically replaced with land use change data based on the simulation with the DLS model. Finally, the data simulated by the regional climate model were matched with meteorological sites, and the meteorological site data were prepared for the simulation with SWAT.

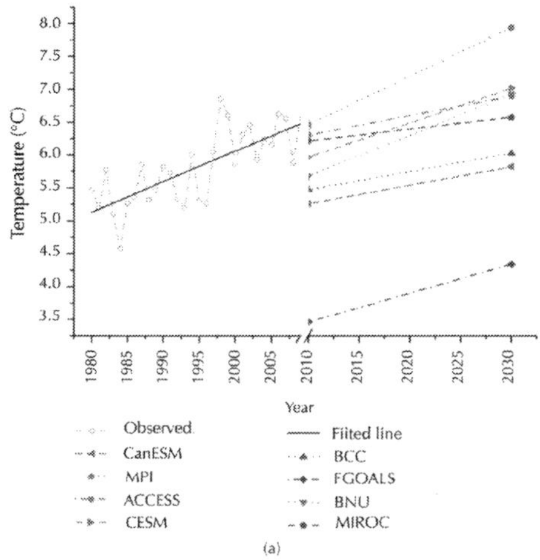

(a)

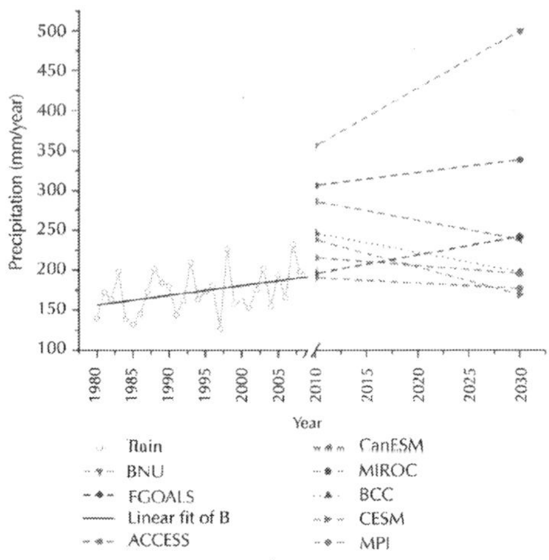

(b)

Figure 2: The trends in (a) observed (grey line; running average in black) and projected (10AR5 GCMs; colored dash lines) temperature, (b) observed and projected precipitation.

Simulation of Hydrological Cycle with the SWAT Model

The study area was first divided into subwatersheds, which were subdivided into hydrological response units (HRUs). Besides, for each subwatershed, the climate data used are taken from the GCM grid point that is the closest to its centroid. To improve performance, the SWAT model was calibrated and validated by adjusting several parameters and comparing the simulated streamflow with observed values. The most sensitive parameters were identified with the built-in sensitivity analysis tool in SWAT [40]. The daily streamflow observation data from Yingluoxia Hydrological Station in 2004 were used for calibration, and the observation data in 2005 were used for validation. It should be noted that the first three years were used as a warm-up period to mitigate the effects of unknown initial conditions, which were then excluded from the subsequent analysis. The ability of the SWAT model to replicate the temporal trends in the historical hydrological observations was assessed using the coefficient of determination (R^2), the Nash and Sutcliffe (1970) model efficiency (NSE), and the root mean square error (RMSE).

RESULT AND DISCUSSION

Calibration and Validation

The SWAT model was calibrated for 2004 and validated for 2005 using the daily streamflow observation data from four gauging stations within the study area. Finally, fifteen parameters were selected for the calibration (Table 2), which are associated with snow (SFTMP, SMTMP, SMFMX, SMFMN, and TIMP), runoff (CN2), groundwater (ALPHA_BF and GW_DELAY), soil (SOL_AWC), channel (CH_N and CH_K2), and evaporation (ESCO) processes. After the sensitivity analysis, 9 relatively more sensitive parameters were identified for the calibration. Most of the parameters were adjusted based on multiple trials, and the SWAT model was calibrated using an automatic calibration technique with the program Sequential Uncertainty Fitting Version (SUFI-2). With

SUFI-2, sensitive initial and default parameters related to hydrology varied simultaneously until an optimal solution was achieved. The most sensitive parameters with their best ranges and best-fitted values are shown in Table 3. Finally, these best-fitted values were used to adjust the initial model inputs for the simulation during 2006–2030. The model was validated using daily streamflow observation data from the Yingluoxia Hydrological Station in 2005. The validation results show that the NSE is 0.78 and R^2 of the observed and simulated data is 0.81 (Figure 3), demonstrating the high behavioral performance of the SWAT model.

Table 3: List of calibration parameters and the optimized values

Parameter	Description	Range	Optimized value
TLAPS	Temperature lapse rate [C/ km]	0, –10	–3.8
PLAPS	Precipitation lapse rate [mmH2O/km]	0, 100	5.8
SFTMP	Snowfall temperature [C]	–2, +2	0.9
SMTMP	Snow melt base temperature [C]	–5, +5	2.1
SNOEB	Initial snow water content in elevation band [mm]	50, 230	100
TIMP	Snowpack temperature lag factor	0.38–0.62	0.49
SMFMN	Melt factor for snow on December 21 [mmH2O/ C-day]	3.05–3.51	3.25
SMFMX	Melt factor for snow on June 21 [mmH2O/ C-day]	5.85–6.27	6.02
SURLAG	Surface runoff lag time [days]	4.18–5.19	4.68

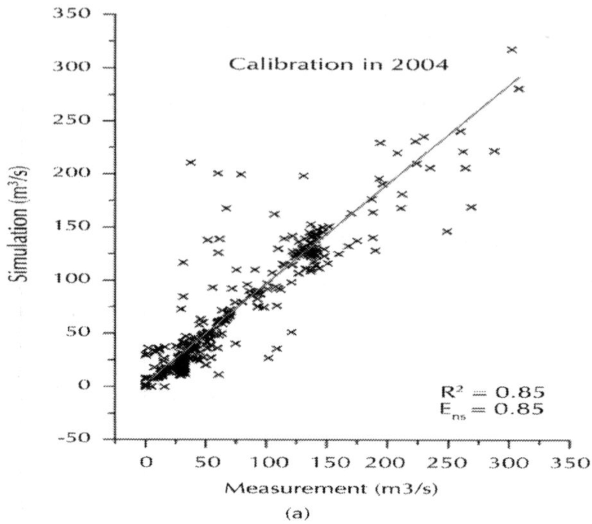

(a)

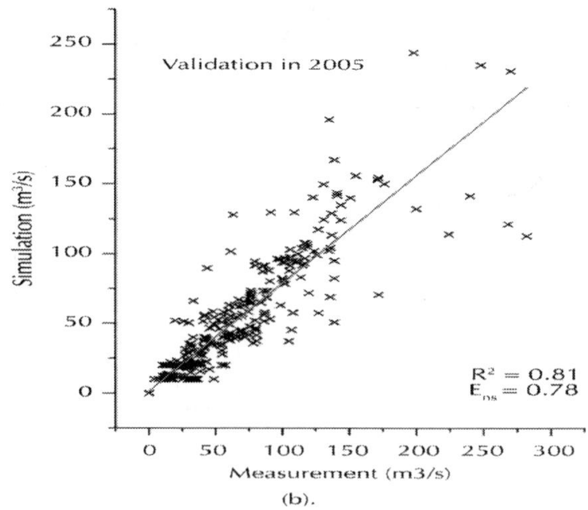

(b)

Figure 3: The calibration and validation on stream flow of the SWAT.

Future Climate under the RCP 4.5 Scenario

Based on the downscaled GCM climate data, we calculated the mean temperature and precipitation of the 9 lattice points around the grid that included Qilian meteorological station. The results were compared with the mean monthly temperatures and precipitation of the meteorological station during 1981–2005. The monthly mean temperatures of the 25 years ranged from −3 to 3°C and increased by around 0.8°C. The mean monthly precipitation ranges from −0.3% to 10% and increased by around 7.8% (Figure 4). The increase range of mean monthly precipitation is large, while the range of reduction is smaller.

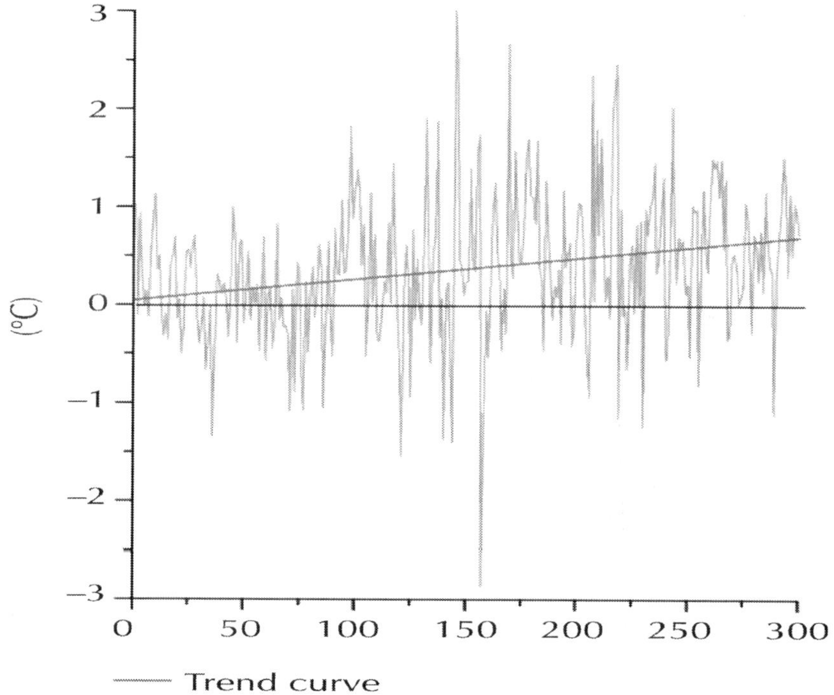

(a)

(a)

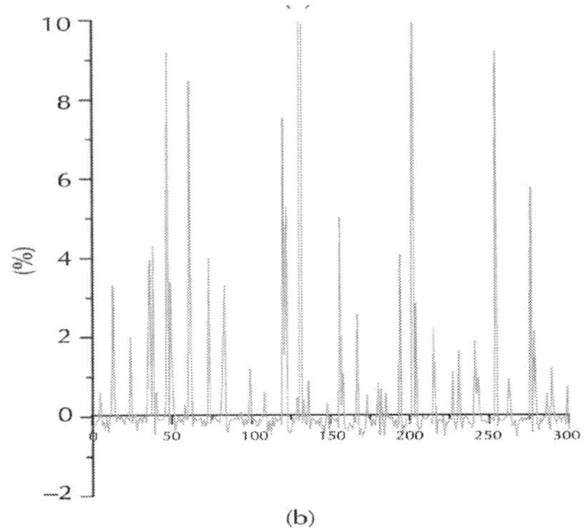

(b)

Figure 4: The difference of average monthly temperature and precipitation between 1981–2005 and 2006–2030 in Qilian Weather observed station.

Future Land Use Change Simulated with DLS

The results suggested the change in one land use type was influenced by multiple factors, and the 13 driving factors can reasonably explain the spatial patterns of all land use types. For example, the existence of forest land was significantly influenced by all the 13 driving factors, while the existence of cultivated land and grassland was affected by the altitude, distance, and soil factors. The future land uses for 2006–2030 were simulated with the DLS model by combining the probability maps prepared with logistic regression analysis, the land demands under different scenarios, and the map of development-restricted areas. The simulation results indicated that the most dramatic land use changes during 2006–2030 will mainly occur in the upper reach and some parts of the middle reach of Heihe River Basin. Compared to 2005, the areas of forest land and unused land in 2030 will decrease by 6.2% and 1.6%, respectively, while the areas of built-up land, cultivated land, and grassland will increase by 1.7%, 1.3%, and 4.8%, respectively (Figure 5). The significant increase of grassland area may mainly result from the steady pasture construction, and this uptrend may continue in the future owing to the increasing demand for pasture products.

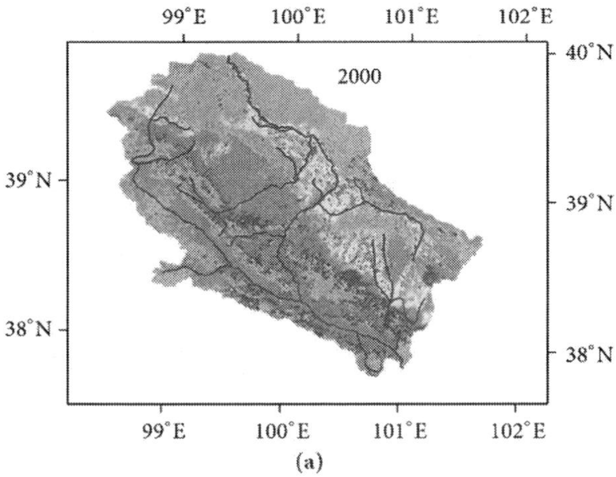

(a)

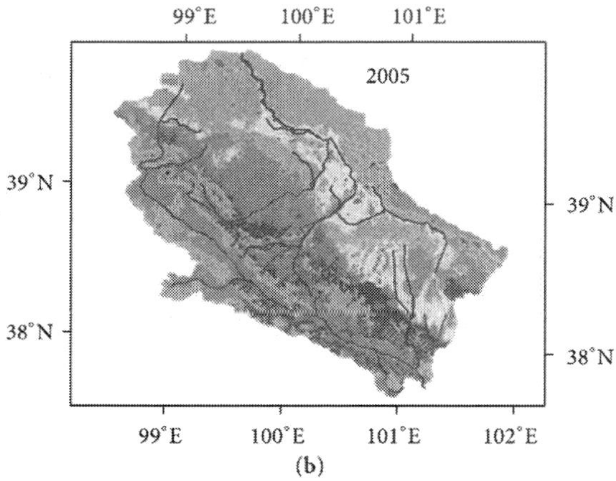

(b)

Figure 5: Comparison of the land uses interpreted in 2000, 2005 and simulated in 2010, 2030.

Impacts of Climate and Land Use Changes on Watershed Hydrology

Four simulation experiments were designed based on the land use data and climate data. In the baseline experiment for the period during 1981–2005, the water yield was simulated with the land use data in 2000, 2005 and the weather station observations during 1981–2005 (Figure 6(b)). Then three scenarios for the period during 2006–2030 were designed based on the land use and climate change (Figure 6(a)), the results from which were compared with that in the baseline experiment. In the first scenario during 2006–2030, the water yield was simulation with the land use data in 2010 and 2030, temperature data during 2006–2030, and the precipitation data during 1981–2005. The simulation result shows that the impacts of future land use change on the water yield vary with seasons, and the land use change will have negative overall influence on the water yield, with an influence degree of −1.8% according to the annual mean water yield.

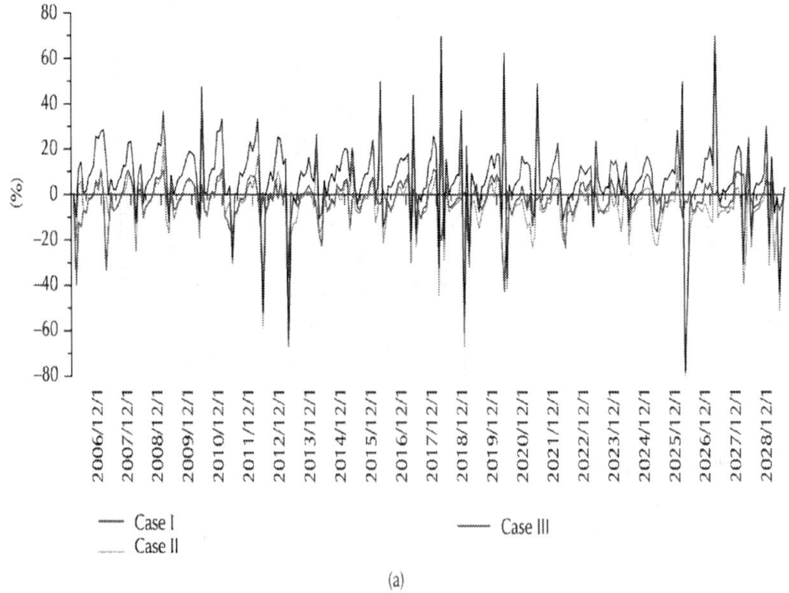

(a)

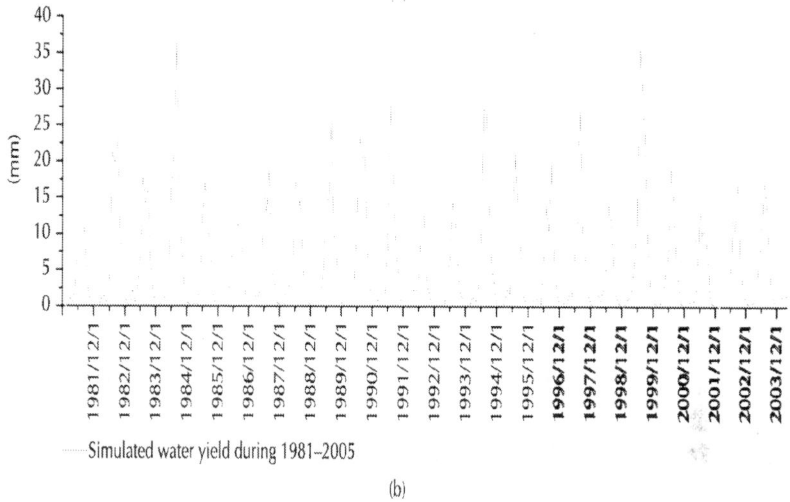

(b)

Figure 6: Comparison of water yields under three scenarios.

The second scenario during 2006–2030 was based on scenarios of temperature and land use changes. The second experiment used the land use data in 2010 and 2030, scenario data of temperature during 2006–2030, and precipitation data during 1981–2005. The analysis of climate change scenarios shows that the average temperature will rise by 0.8°C between 1981–2005 and 2006–2030. The simulation result in the second experiment shows that the land use and temperature changes will make the water yield change by 0.6%–1.1%, the change range of which is relatively smaller compared to the simulation results under the scenario with only land use change. The reasons may be that the temperature rise and melting of a small amount of snow slightly offset the adverse effects of land use change. At the same time, the higher temperatures will result in more winter precipitation in the form of rain rather than snow, leading to the hydrologic consequences including increased winter discharge, a shift in the spring snowmelt peak to earlier in the season, and decreased summer discharge.

The third scenario during 2006–2030 involves scenarios of changes in all of land use, temperature, and precipitation. The land use data in 2010 and 2030 and temperature and precipitation data during 2006–2030 were used under the third scenario. The simulation result shows that these three factors jointly have positive impacts on the water yield, making the basin water yield increase by about 9.8%. The increase of the basin water yield is mainly caused by the change in precipitation, which will increase by around 10.8% during 2006–2030 in comparison to that during 1981–2005. Overall, the simulation results suggest that the basin water yield will increase in the future under different scenarios of climate and land use changes.

DISCUSSION AND CONCLUSIONS

In the Heihe River Basin, the upper reaches are featured with the generation and use of blue water, while the lower reach and surrounding areas are characterized by natural ecosystems and a low population density. LULC is defined as syndromes of human activities such as agriculture, forestry, and building construction, and most of previous studies only focused on the hydrological influence of LULC change in the upper reach. The separation between hydrological impacts of land use and climate changes has never been studied in the upper and middle reaches of the Heihe River Basin. However, we argue that studying the hydrological processes in the upper and middle reaches is essential since water supply to the lower reach is impacted by both the climate change and human activities in the upper and middle reaches.

In this study, we analyzed the impacts of potential climate and land use changes on the water yield in the upper and middle reaches of Heihe River Basin based on the simulation with the SWAT model. The results show that the water yield was more affected by climate change than by land use change. This indicates that the predicted increase in precipitation will exert more significant impacts on the watershed hydrology than the predicted land use changes will. However, the analysis of the projected streamflow changes shows that there are higher uncertainties in the dry season compared with the wet season in the simulation with the hydrological model and GCMs climate data. It is difficult to accurately project the hydrological changes since there are various uncertainties associated with the future Green House Gas (GHG) emission scenarios, GCM structure, downscaling method, LULC, and hydrological models. In particular, water resource managers are generally confronted with complex problems in sustainable management and conservation of water resources due to the uncertainties in the future hydrological projection under climate and land use changes. It is therefore crucial to consider both land use and climate changes in water resource planning for the Heihe River Basin so as to mitigate their negative hydrological impacts, and more valuable information may be provided to the water resource managers if these uncertainties in the future hydrological projection can be effectively reduced through advanced modeling and research.

ACKNOWLEDGMENTS

This research was financially supported by the major research plan of the National Natural Science Foundation of China (Grant no. 91325302), National Basic Research Program of China (973 Program) (no. 2010CB950904), and the National Natural Science Funds of China for Distinguished Young Scholar (Grant no. 71225005).

REFERENCES

1. Z. Žalud, M. Trnka, M. Dubrovský, P. Hlavinka, D. Semerádová, and E. Kocmánková, "Climate change impacts on selected aspects of the Czech agricultural production," Plant Protection Science, vol. 45, pp. S11–S19, 2009.
2. E. Lorencová, J. Frélichová, E. Nelson, and D. Va ká , "Past and future impacts of land use and climate change on agricultural ecosystem services in the Czech Republic," Land Use Policy, vol. 33, pp. 183–194, 2013.
3. R. Mahmood, A. I. Quintanar, G. Conner, et al., "Impacts of land use/land cover change on climate and future research priorities," Bulletin of the American Meteorological Society, vol. 91, no. 1, pp. 37–46, 2010.
4. Z. Ma, S. Kang, L. Zhang, L. Tong, and X. Su, "Analysis of impacts of climate variability and human activity on streamflow for a river basin in arid region of northwest China," Journal of Hydrology, vol. 352, no. 3-4, pp. 239–249, 2008.

5. B. R. Scanlon, I. Jolly, M. Sophocleous, and L. Zhang, "Global impacts of conversions from natural to agricultural ecosystems on water resources: quantity versus quality," Water Resources Research, vol. 43, no. 3, Article ID W03437, 2007.

6. L. M. Mango, A. M. Melesse, M. E. McClain, D. Gann, and S. G. Setegn, "Land use and climate change impacts on the hydrology of the upper Mara River Basin, Kenya: results of a modeling study to support better resource management," Hydrology and Earth System Sciences, vol. 15, no. 7, pp. 2245–2258, 2011.

7. C. J. Vörösmarty, P. Green, J. Salisbury, and R. B. Lammers, "Global water resources: vulnerability from climate change and population growth," Science, vol. 289, no. 5477, pp. 284–288, 2000.

8. S. Qi, G. Sun, Y. Wang, S. G. McNulty, and J. A. M. Myers, "Streamflow response to climate and landuse changes in a coastal watershed in North Carolina," Transactions of the ASABE, vol. 52, no. 3, pp. 739–749, 2009.

9. J. Kim, J. Choi, C. Choi, and S. Park, "Impacts of changes in climate and land use/land cover under IPCC RCP scenarios on streamflow in the Hoeya River Basin, Korea," Science of the Total Environment, vol. 452-453, pp. 181–195, 2013.

10. Z. Li, W.-Z. Liu, X.-C. Zhang, and F.-L. Zheng, "Impacts of land use change and climate variability on hydrology in an agricultural catchment on the Loess Plateau of China," Journal of Hydrology, vol. 377, no. 1-2, pp. 35–42, 2009.

11. R. W. Skaggs, D. M. Amatya, G. Chescheir, C. Blanton, and J. Gilliam, "Effect of drainage and management practices on hydrology of pine plantation," in Proceedings of the International Conference on Hydrology and Management of Forested Wetlands, 2006.

12. T. D. Prowse, S. Beltaos, J. T. Gardner et al., "Climate change, flow regulation and land-use effects on the hydrology of the Peace-Athabasca-Slave system; Findings from the Northern Rivers Ecosystem Initiative," Environmental Monitoring and Assessment, vol. 113, no. 1–3, pp. 167–197, 2006.

13. B. Dixon and J. Earls, "Effects of urbanization on streamflow using SWAT with real and simulated meteorological data," Applied Geography, vol. 35, no. 1-2, pp. 174–190, 2012.

14. E. Lioubimtseva, R. Cole, J. M. Adams, and G. Kapustin, "Impacts of climate and land-cover changes in arid lands of Central Asia," Journal of Arid Environments, vol. 62, no. 2, pp. 285–308, 2005.
15. Z. Wang, D. L. Ficklin, Y. Zhang, and M. Zhang, "Impact of climate change on streamflow in the arid Shiyang River Basin of northwest China," Hydrological Processes, vol. 26, no. 18, pp. 2733–2744, 2012.
16. D. R. D›Agostino, L. G. Trisorio, N. Lamaddalena, and R. Ragab, "Assessing the results of scenarios of climate and land use changes on the hydrology of an Italian catchment: modelling study," Hydrological Processes, vol. 24, no. 19, pp. 2693–2704, 2010.
17. V. Mishra, K. A. Cherkauer, D. Niyogi, et al., "A regional scale assessment of land use/land cover and climatic changes on water and energy cycle in the upper Midwest United States," International Journal of Climatology, vol. 30, no. 13, pp. 2025–2044, 2010.
18. J. A. Vano, J. A. Foley, C. J. Kucharik, and M. T. Coe, "Evaluating the seasonal and interannual variations in water balance in Northern Wisconsin using a land surface model," Journal of Geophysical Research G: Biogeosciences, vol. 111, no. 2, 2006.
19. D. Mao and K. A. Cherkauer, "Impacts of land-use change on hydrologic responses in the Great Lakes region," Journal of Hydrology, vol. 374, no. 1-2, pp. 71–82, 2009.
20. J.-Y. Park, M.-J. Park, M.-J. H.-K. Joh et al., "Assessment of MIROC3.2 hires climate and clue-s land use change impacts on watershed hydrology using Swat," Transactions of the ASABE, vol. 54, no. 5, pp. 1713–1724, 2011.
21. S. J. Kim, H. J. Kwon, G. A. Park, and M. S. Lee, "Assessment of land-use impact on streamflow via a grid-based modelling approach including paddy fields," Hydrological Processes, vol. 19, no. 19, pp. 3801–3817, 2005.
22. E. Lu, E. S. Takle, and J. Manoj, "The relationships between climatic and hydrological changes in the upper Mississippi river basin: a SWAT and multi-GCM study," Journal of Hydrometeorology, vol. 11, no. 2, pp. 437–451, 2010.
23. L. Liu, Z. Liu, X. Ren, T. Fischer, and Y. Xu, "Hydrological impacts of climate change in the Yellow River Basin for the 21st century

using hydrological model and statistical downscaling model," Quaternary International, vol. 244, no. 2, pp. 211–220, 2011.

24. R. B. Singh and C. Shi, "Advances in observation and estimation of land use impacts on climate changes: Improved data, upgraded models, and case studies," Advances in Meteorology, vol. 2014, Article ID 748169, 7 pages, 2014.
25. S. Fall, D. Niyogi, A. Gluhovsky, R. A. Pielke, E. Kalnay, and G. Rochon, "Impacts of land use land cover on temperature trends over the continental United States: assessment using the North American Regional Reanalysis," International Journal of Climatology, vol. 30, no. 13, pp. 1980–1993, 2010.
26. H. Guo, Q. Hu, and T. Jiang, "Annual and seasonal streamflow responses to climate and land-cover changes in the Poyang Lake basin, China," Journal of Hydrology, vol. 355, no. 1–4, pp. 106–122, 2008.
27. L. Tang, D. Yang, H. Hu, and B. Gao, "Detecting the effect of land-use change on streamflow, sediment and nutrient losses by distributed hydrological simulation," Journal of Hydrology, vol. 409, no. 1-2, pp. 172–182, 2011.
28. M. D. Tomer and K. E. Schilling, "A simple approach to distinguish land-use and climate-change effects on watershed hydrology," Journal of Hydrology, vol. 376, no. 1-2, pp. 24–33, 2009.
29. P. F. Juckem, R. J. Hunt, M. P. Anderson, and D. M. Robertson, "Effects of climate and land management change on streamflow in the driftless area of Wisconsin," Journal of Hydrology, vol. 355, no. 1–4, pp. 123–130, 2008.
30. Q. Feng, G. D. Cheng, and K. N. Endo, "Towards sustainable development of the environmentally degraded River Heihe basin, China," Hydrological Sciences Journal, vol. 46, no. 5, pp. 647–658, 2001.
31. G.-D. Cheng, H.-L. Xiao, Z.-M. Xu, J.-X. Li, and M.-F. Lu, "Water issue and its countermeasure in the inland river basins of Northwest China—a case study in Heihe River Basin," Journal of Glaciology and Geocryology, vol. 28, no. 3, pp. 406–413, 2006.
32. A. Jarvis, H. I. Reuter, A. Nelson, and E. Guevara, Hole-Filled Seamless SRTM Data V3, International Centre for Tropical Agriculture (CIAT), 2006.

33. X. Deng, Q. Jiang, H. Su, and F. Wu, "Trace forest conversions in Northeast China with a 1-km area percentage data model," Journal of Applied Remote Sensing, vol. 4, no. 1, Article ID 041893, pp. 1–13, 2010.
34. X. Deng, H. Su, and J. Zhan, "Integration of multiple data sources to simulate the dynamics of land systems," Sensors, vol. 8, no. 2, pp. 620–634, 2008.
35. X. Deng, J. Liu, Y. Lin, and C. Shi, "A framework for the land use change dynamics model compatible with rcms," Advances in Meteorology, vol. 2013, Article ID 658941, 7 pages, 2013.
36. X. Deng, C. Zhao, and H. Yan, "Systematic modeling of impacts of land use and land cover changes on regional climate: a review," Advances in Meteorology, vol. 2013, Article ID 317678, 11 pages, 2013.
37. X. Deng, F. Yin, Y. Lin, Q. Jin, and R. Qu, "Equilibrium analyses on structural changes of land uses in Jiangxi Province," Journal of Food, Agriculture and Environment, vol. 10, no. 1, pp. 846–852, 2012.
38. X. Deng, Q. Jiang, J. Zhan, S. He, and Y. Lin, "Simulation on the dynamics of forest area changes in Northeast China," Journal of Geographical Sciences, vol. 20, no. 4, pp. 495–509, 2010.
39. X. Deng, C. Zhao, Y. Lin et al., "Downscaling the impacts of large-scale LUCC on surface temperature along with IPCC RCPs: a global perspective," Energies, vol. 7, no. 4, pp. 2720–2739, 2014.
40. S. Neitsch, J. Arnold, J. E. A. Kiniry, R. Srinivasan, and J. Williams, "Soil and water assessment tool user›s manual version 2000," GSWRL Report 02-06, 2002.

Chapter 5

Systematic Modeling of Impacts of Land Use and Land Cover Changes on Regional Climate: A Review

Xiangzheng Deng[1,2], Chunhong Zhao[1,2,3], and Haiming Yan[4]

[1]Institute of Geographic and Natural Resources Research, Chinese Academy of Sciences, Beijing 100101, China

[2]Center for Chinese Agricultural Policy, Chinese Academy of Sciences, Beijing 100101, China

[3]University of Chinese Academy of Sciences, Beijing 100049, China

[4]State Key Laboratory of Water Environment Simulation, School of Environment, Beijing Normal University, Beijing 100875, China

ABSTRACT

There have been tremendous changes in the global land use pattern in the past 50 years, which has directly or indirectly exerted significant influence on the global climate change. Quantitative analysis for the impacts of land use and land cover changes (LUCC) on surface climate is one of the core scientific issues to quantitatively analyze the impacts of LUCC on the climate so as to scientifically understand the influence of human activities on the climate change. This paper comprehensively analyzed the primary scientific issues about the impacts of LUCC on the regional climate and reviewed the progress in relevant researches. Firstly, it introduced the influence mechanism of LUCC on the regional climate and reviewed the progress in the researches on the biogeophysical process and biogeochemical process. Then the model simulation of effects of LUCC on the regional climate was introduced, and the development from the global climate model to the regional climate model and the integration of the improved land surface model and the regional climate model were reviewed in detail. Finally, this paper discussed the application of the regional climate models in the development and management of agricultural land and urban land.

INTRODUCTION

The land use and cover changes (LUCC), in which the human activities play a dominant role, interact with the environment and have significant effects on the ecosystems at the local, regional, and global scales and consequently directly or indirectly exert great influence on the global climate changes [1–4]. People have come to realize that the global change always consists of a series of regional changes with various processes and patterns according to the more and more observations [5]. On the one hand, the initial regional change will gradually extend to the interregional and even larger scales and consequently influence the change of the global environment. So it is necessary to first take into account the local and regional climatic effects in the research on the effects of the LUCC on the climate. On the other hand, since there is a great difference in the physical, chemical, and biological characteristics of the land surface in different regions on the Earth, the impacts of the land use and land cover changes on the regional climate systems also vary greatly. For example, the deforestation may lead to the temperature increase in the tropic zone, while it may lead to the regional temperature decrease in the frigid zone [6]. Therefore, in order to completely reflect the relationship between LUCC and the climatic factors, it is necessary to carry out and make comparisons among a lot of case studies in different regions and at various spatiotemporal scales and analyze the spatiotemporal processes and land surface parameters of the LUCC and the response of the climate. However, there have been very few relevant researches.

It has long been known that the land use changes caused by human activities such as deforestation and agriculture practice have some effects on the climate [1, 7]. The land use changes can influence climate by changing the properties of the land surface which is not only the direct heat source of the troposphere, but also one of the main sources of the atmospheric vapor water [8]. Therefore, the change in the characteristics of the land surface will directly influence the land surface-atmosphere interaction and consequently alter the thermodynamic and dynamic characteristics of the atmosphere and finally lead to different climate processes and patterns [9]. The land use activities have significantly changed the regional land cover, thus leading to climate changes. For example, the deforestation [10], afforestation [11], agriculture practice [7], and urbanization [12, 13] all influence the energy budget and Bowen ratio of the land surface, the distribution of the precipitation among the soil water, runoff, and evapotranspiration. The land use activities also have important influence on the characteristics of regional climate system, for example, the temperature, evapotranspiration, precipitation, wind field, atmospheric pressure, and especially the temperature and precipitation [14–16]. In addition, the recent researches suggest that the LUCC may affect the extremes in temperature [2, 17] and precipitation [18].

The LUCC affects the local, regional, and global climate systems through various biogeochemical and biogeophysical processes [15, 19, 20]. For example, the biogeochemical process can indirectly affect the climate by altering the rate of the biogeochemical cycle and thereby changing the chemical composition of the atmosphere [21]. Besides, it may also affect the climate through the absorption or emission of greenhouse gases [21]. By contrast, the biogeophysical process directly affects the physical parameters that determine the absorption and disposition of energy at the land surface [15]. For example, it influences albedo or reflective properties of the land surface [22], alters the absorption rate of solar radiation, and hence influences the energy availability at the land surface [23]. In addition, the surface hydrology and vegetation transpiration characteristics also affect how the energy received by the surface is distributed into the latent and sensible heat fluxes [4]. The vegetation structure can affect the surface roughness and thereby alter the momentum and heat transport.

Owing to the limitation of knowledge and lack of interdisciplinary cooperation, there is still lack of full understanding of the mechanism of the procedures of the climate process and how the land use changes influence the regional climate and consequently influence the global climate [4]. There are separate studies of the biogeochemical and biogeophysical effects of the land use changes on the climate in their respective fields, and there are few researches on their joint effects on the climate. However, there are still a lot of uncertainties in the current knowledge, and it is still very difficult to objectively assess the contribution of the land use changes to the climate change [15]. For example, Hansen et al. [24] proposed that the radiative forcing due to the change in albedo caused by the global land use changes may decrease the temperature. By contrast, Intergovernmental Panel on Climate Change (IPCC) believed that the increase of CO_2 in the atmosphere will reinforce the radiative forcing, 1/3 of which was contributed by the land use changes, and consequently increase the temperature [2]. So it is of great importance to objectively analyze the biogeochemical and biogeophysical effects of the land use changes on the climate [1]. It is necessary to carry out further researches on the development of more accurate monitoring and modeling methods, which are essential to the understanding of these interactions and feedbacks [25]. However, the major challenge is how to integrate the researches on the biogeophysical and biogeochemical processes.

The impacts of land use changes on the climate are the synthetic effects of the biogeophysical process and biogeochemical process. But which of the two kinds of processes makes a greater contribution to the climate change at the regional scale, or which one plays a dominant role? The climate model is an effective tool in the study of climate, but how should it be applied to the research on the regional effects of LUCC? What effects will the improvement of the regional climate model have on the LUCC simulation? This paper reviewed the history and methods of the relevant researches, and summarized the influence of LUCC on the regional climate system and the simulation strategies according to the researches in the recent decades. Finally, the application of the regional climate models in the development and management of agricultural land and urban land was discussed.

MECHANISM RESEARCH ON THE INFLUENCE OF LUCC ON THE REGIONAL CLIMATE

The research on the influence of the land use changes on the regional temperature and precipitation can be dated back to the 1970s. The land use changes can alter various physical characteristics of the land surface, including the land surface parameters such as albedo [22], upward long-wave radiation, and roughness and the vegetation parameters such as the vegetation coverage and leaf area index (LAI) [26]. All the changes in these parameters can influence the energy budget and water budget through the land-surface process, have an effect on the atmospheric boundary layer, and further impact the free air [27]. So it can be said that the changes in the parameters of the land surface and vegetation are the fundamental reason for the effects of the land use changes on the atmosphere [28]. By the 1980s, the researchers realized that the land use influences the climate not only through the change in physical characteristics of the land surface, but also through the greenhouse emission. Thereafter, there were more and more researches on the effects of land use on the carbon cycle. The influence of the land use changes on the climate involves not only the biogeophysical process, but also the biogeochemical process [20].

Studies on Influence of Physical Characteristics on the Climate

Charney et al. [29] first explored the relationship between the change of albedo and the drought in the Sahara region. Thereafter, there were a lot of researches on the influence on the regional climate (e.g., temperature and precipitation), which is exerted by the single parameter of physical characteristics of the land surface (e.g., albedo, roughness, leaf area index, and soil moisture) [30, 31]. The change in the underlying landscape alters the energy and moisture budgets of the land surface at the regional scale, which in turn leads to the changes in the fluxes of heat, water, and dynamics of the near-surface atmosphere and influences the key thermodynamic and dynamic properties of the air, which is of great importance to the air convection [32]. The evidence to support this statement comes from the sensitivity studies that explore the impacts of the change on the characteristics of land surface [33, 34].

There is considerable evidence of the significance of land surface processes generated through regional-scale perturbation experiments and researches on the deforestation [6, 35, 36], desertification [37, 38], and land use changes [4, 39]. These researches all indicated the LUCC contributed to the large and statistically significant change in the temperature, rainfall, and other variables at the continental or regional scales.

First and foremost, some important researches on the sensitivity of climate to the change of land surface evapotranspiration were performed by Claussen et al. [40] and Davin and de Noblet-Ducoudre [41], which provided significant evidence that the change of the land surface evapotranspiration can lead to great changes in the temperature and precipitation. For example, the deforestation will lead to the decrease of evapotranspiration [39], while the decrease of latent heat will increase the near-surface temperature and lead to higher sensible heat flux [41]. There is a high transpiration rate in the tropic rainforest, where the decrease of the transpiration rate due to the disappearance of the tropic rainforest may lead to regional warming and drought in the future [4, 42]. In addition, the change in evapotranspiration also influences the water content in the atmosphere and reduces the greenhouse effect and consequently reduces the temperature [34], while the decrease of the cloud covers will increase the solar radiation and consequently strengthen the temperature [43]. Therefore, due to the large number of parameters, it is difficult to quantitatively compare the two feedbacks and there is great uncertainty [2].

The land surface albedo is the fraction of solar energy (shortwave radiation) reflected from the land surface into space. It represents the reflecting power of the land surface and plays a key role in influencing the radiation balance and energy balance of the land surface [44]. The change in the land surface albedo directly alters the solar radiation absorbed by the land surface, subsequently leads to the change in the long-wave radiation of the land surface into space and the sensible heat and latent heat, and finally influences the temperature [45]. Some of the key works demonstrating the sensitivity of climate to land surface albedo were performed by Charney et al. [29]. Without regard to the influence of the advection process, the increase will lead to the decrease of the solar radiation absorbed by the land surface, increase of the land surface temperature, subsequently increase of the long-wave radiation into the space and decrease the sensible heat and latent heat, and lead to the potential to decrease the temperature, and vice versa [45]. The lower land surface albedo indicates the lower reflectivity and higher shortwave absorption of the land surface. Taking the forest coverage as an example, its albedo is generally lower than that of the nudation and other vegetation types [46]. The historical deforestation in the middle-latitude zone led to the increase of the land surface albedo, which is especially remarkable after snowfall in the winter and might have made the northern hemisphere colder [4].

The leaf area index is an important indicator that represents the canopy structure and productivity of the plant community. It directly influences the ability of the plant to acquire and utilize the solar energy and indirectly influences the canopy impedance ratio. Besides, being an important parameter of the land surface albedo, it also directly influences the interaction between the land surface and the atmosphere [47]. More recent studies have focused on the role of the leaf area index in influencing the climate. The relevant researches indicate that the leaf area index is closely related to the precipitation, temperature, specific humidity, and so forth [48].

The significance of roughness has also been well realized [49]. The land surface roughness has impacts on the turbulent flow between the land surface and the atmosphere and subsequently influences the local diffusion flux. If the diffusion flux is higher, it will reduce the near-surface air temperature under the condition of no other feedback mechanisms [25]. The deforestation will decrease the land surface roughness and reduce the turbulent flow and will consequently increase the temperature in theory. However, the less turbulent flow will lead to the decrease of the heat and moisture transfer, which will increase the temperature and moisture gradients between the land surface and the atmosphere and in turn alleviate the warming effect [49]. Davin and de Noblet-Ducoudre [41] analyzed the sensitivity to the roughness during the conversion of the forest to the grassland. Their result indicates that the change in the roughness will make the global temperature increase by 0.29°C, and the increase will be even more obvious in the tropic zone.

Impacts on the Biogeochemical Process

The influence of the land use change on the biogeochemical process, especially the discharge or absorption of the greenhouse gases such as CO_2 in the atmosphere due to the land use change, can alter the concentration of the greenhouse gases in the atmosphere and consequently influence the climate [13]. The historical accumulative carbon loss due to the land use change was estimated to be 180–200 PgC [50], and the land use change contributed to 10%–30% of the carbon discharge due to human activities. The deforestation, afforestation, forest restoration, and agricultural activities are the major approaches through which the land use influences the carbon cycle. There are many researches on the influence of the land use change on the carbon cycle, most of which focused on the deforestation, especially the deforestation in the tropic rainforest. Since the 1850s, the global forest area decreased by 20%, and the carbon emission due to the deforestation accounted for 90% of the carbon emission caused by the land use change [51] and 33% of the man-made carbon emission (including the discharge from the fossil fuel burning and land use change) [4].

The influence of the afforestation and forest restoration on the carbon cycle has also gradually become the hot issue in the relevant researches. Although the reforestation has not significantly influenced the terrestrial carbon sink on the global scale, it has played an important role in the carbon sink on the regional scale. For example, the man-made forests in China have stored over 0.45 Gt carbon since the 1970s [52]. Some researches indicated the forest restoration played a key role in the carbon sink resulting from the land use [53, 54].

The productivity of the forest and decomposition of organic matters in the soil both influence the change of CO_2 in the atmosphere and the climate pattern and consequently influence the terrestrial carbon sink. When the photosynthesis rate is higher than the respiration rate, discharge rate of the biogenic volatile organic compounds (BVOC), and the decomposition rate of organic carbon, the forest plays a role as the carbon sink [55]. Besides, the deforestation will reduce the potential carbon sequestration [50]. In addition, the higher temperature and higher CO_2 concentration will increase the NPP, which will lead to the negative feedback and make more CO_2 in the atmosphere sequestrated.

The radiative forcing due to the carbon emission resulting and the accelerated respiration due to the higher temperature will lead to the positive feedback, in which the organic matters will decompose more rapidly [56]. For example, there is an extreme example in the climate-carbon cycle; that is, the tropic rainforest in Amazon will gradually succeed into another vegetation if the temperature continuously increases and the precipitation continuously decreases.

The agricultural activities also have significant impacts on the carbon cycle. For example, the conversion of the natural vegetation to the cultivated land, loss of plant biomass, and increased decomposition of organic matters in the soil all make the agricultural activities become one of the major sources of CO_2 emission. By contrast, the utilization of the high yield variety and fertilizers, irrigation, and no-till agriculture all contribute to the reduction of carbon loss and increase the absorption of carbon in the agricultural regions [56]. For example, the no-till agriculture in the USA has increased the organic content in the soil, reaching 1.4 Gt carbon in the past 30 years. However, the increase of the organic matters in the soil can only last for 50–100 years, after which a new equilibrium of carbon cycle will be reached [57].

A lot of BVOCs are generated by the plants and then released into the atmosphere. These compounds have significant influence on the physical and chemical characteristics of the atmosphere due to their great amount [58]. Although the BVOCs have no direct influence on the radiation balance of the atmosphere, they affect the longevity of the methane in the atmosphere and play an important role in the formulation of the ozone and secondary organic aerosols (SOAs). The SOAs can directly influence the climate since they can scatter or absorb the solar radiation and consequently decrease or increase the temperature [59]. The overall influence of the SOAs on the climate system has not been accurately quantified so far; they may mainly play a role in decreasing the temperature [45]. The SOAs also have significant indirect impacts on the climate; that is, they may act as the cloud condensation nuclei (CCN) in the formation of the clouds. In fact, the BVOCs released by the boreal forest have made the local CCN increase by 100% [59], which in turn influences the number of water drops in the cloud and makes the albedo of the atmosphere increase by 3%–8%.

There will be an albedo difference ranging from −1.8 W/m^2 to −6.7 W/m^2 in the regions between 60° and 90° N since the albedo increases due to the indirect influence of BVOC. This means that, owing to the feedback among BVOCs released by the trees, SOA, CNN, and land surface albedo, the boreal forest will make the local climate colder. The influence of BVOC on CCN is considered to be the most important in the boreal forest since the regional air pollution is slight [59]. In addition, the net emission of carbon released due to the deforestation depends on the land use type converted from the forest or the temporal scale of the regeneration of the forest and the feedback mechanism mentioned previously. According to the simulation which only takes into account the biogeochemical influence, the complete deforestation will make the global temperature increase by 0. 09°C or 0.19°C if the forests between 50–60° N and 0–10° S were completely felled [45]. This is because of the great amount of biomass in the tropic zone and depends on the total biomass of different forest types.

The impacts of the LUCC on the regional temperature and precipitation are the synthetic effects of the processes mentioned previously. There has been great progress in the research on the single process and the synthetic effects. Besides, the impacts of the land cover change happen at certain temporal and spatial scales. For example, the mesoscale land cover change influences the regional temperature through the land surface albedo in short and medium terms, while the geochemical process influences the change in temperature at larger temporal and spatial scales.

SIMULATION OF THE EFFECTS OF LUCC ON THE REGIONAL CLIMATE MODEL

Development from the Global Climate Models to the Regional Climate Models

The climate model consists of the set of equations that can be used to determine the characteristics and evolution of the components of the climate system according to the fundamental physical laws such as the law of conservation of energy and law of conservation of mass. Then the climate model is constructed by routinizing the set of equations with the computer. The climate model can be used to not only simulate the current climate, but also to simulate the climate change caused by the change in the boundary conditions [60]. Therefore, the climate model will serve as the most important test tool if the people wish to study the climate and its change with the experimental method.

The earliest researches on the regional climate effects generally used the general circulation models (GCMs), and carried out the sensitivity test with the force-response method [61, 62], that is, represent the land use changes with the changes of the land surface parameters (e.g., albedo, roughness, and evapotranspiration). Generally, the control experiment is first carried out with the GCMs, and then the land surface parameters are changed to carry out the simulation and thereafter compare with the reference test so as to analyze the response of the climate to the land use changes [48]. Charney et al. [29] first applied this method in the study of the positive feedback mechanism between the land surface albedo and aridification in the Sahara Desert. Thereafter, the method was applied to many researches on the response of climate to the change of the land surface parameters such as the evapotranspiration [63], roughness [49], stomatal conductance [64], and leaf area index [65].

The GCMs have been widely used in the study of climate change, and a lot of effective work has been done in the simulation of influence of the LUCC on the global temperature and precipitation at large scale and the research on the climate effects of trace gases in the atmosphere and so forth [13, 66, 67]. However, it is difficult to simulate the regional climate at small scales precisely with the GCMs due to the limitation of the resolution. Besides, the resolution is very low in the GCMs (above 100–200 km), so they cannot be used to describe the complex terrain and land surface characteristics. Therefore, there are some bias and uncertainties in the simulation of the regional climate change with the GCMs [68], which influences the credibility of the simulation result.

The regional climate model was proposed during the late 1980s and the early 1990s [69] and has become an important tool in the regional climate researches [70]. The regional climate model has higher resolution and can more reasonably simulate the regional forcing such as the terrain, rivers/lakes, and urban buildings and describe the detailed characteristics of the land surface. Besides, it can reflect the climate characteristics caused by the strong regional forcing [71], thus consequently having been widely used in the regional climate researches.

More attention has been paid to the response of regional and local climate to the land use change rather than that of the global climate. However, the resolution of the current climate models, for example, the global ocean-atmosphere-land system with high complexity and the Earth system model of the moderate complexity, is generally too low when the model is used at the local and regional scales. In the recent decades, the high resolution regional climate models, for example, RegCM2, RegCM3, RAMS, RIEMS, RegCM-NCC, and IPCR-RegCM, have been widely used in the research on the regional land use change. For example, these models have been applied to the desertification experiment [72], deforestation experiment in the tropic rainforest [73], vegetation restoration experiment [32], and so forth. Besides, some researchers analyze the impacts of land use change on the regional temperature and precipitation and so forth [7] and the influence of land use changes on the fundamental regional climate characteristics at the national scale [48, 74, 75] by comparing the current land use data and the data of potential natural vegetation.

Research on the Integration of the Improved Land Surface Model and Regional Climate Model

Although some regional climate models have higher resolution (below 100 km), there are still a lot of great challenges in the simulation of the land surface processes. For example, it is still necessary to represent the land surface process which is very important to the climate at the small scales, although there is little evidence that the simulation with climate models can provide reliable information at small scales [76, 77]. Since the atmosphere and the land surface integrate into an inseparable whole system through the exchange of energy, dynamics and, moisture, it is the key to the successful simulation with the regional model to construct a land surface model that can accurately

As for the intersection between the researches on the land use changes and the climate change, the researches on the climate effects of the future land use changes mainly focus on two important issues. First, great efforts have been made to understand the climate effects of the past and current LUCC; however, there have been very few researches on that of the future LUCC. Second, how to predict the land use changes in the future and predict and assess its influence on the future climate may become a hot issue in the relevant research fields.

ACKNOWLEDGMENTS

This research was financially supported by the National Key Programme for Developing Basic Science in China (Grant no. 2010CB950904), China National Natural Science Funds for Distinguished Young Scholar (Grant no. 71225005), and the Project of the National Natural Science Foundation of China (Grant no. 41071343).

REFERENCES

1. J. A. Foley, R. DeFries, G. P. Asner et al., "Global consequences of land use," Science, vol. 309, no. 5734, pp. 570–574, 2005.
2. R. Pachauri and A. Reisinger, IPCC Fourth Assessment Report, Intergovernmental Panel on Climate Change (IPCC), Geneva, Switzerland, 2007.
3. D. Ojima, Global Land Project: Science Plan and Implementation Strategy, IGBP Secretariat, 2005.
4. J. J. Feddema, K. W. Oleson, G. B. Bonan et al., "Atmospheric science: the importance of land-cover change in simulating future climates," Science, vol. 310, no. 5754, pp. 1674–1678, 2005.
5. C. Fu, W. Dong, G. Wen, and D. Ye, "Regional response and adaptation to global change," Acta Meteolorogica Sinica, vol. 61, no. 2, pp. 245–249, 2003.
6. G. B. Bonan, "Forests and climate change: forcings, feedbacks, and the climate benefits of forests,"Science, vol. 320, no. 5882, pp. 1444–1449, 2008.

7. L. M. Kueppers and M. A. Snyder, "Influence of irrigated agriculture on diurnal surface energy and water fluxes, surface climate, and atmospheric circulation in California," Climate Dynamics, vol. 38, no. 5-6, pp. 1017–1029, 2012.
8. A. K. Betts, J. H. Ball, A. C. M. Beljaars, M. J. Miller, and P. A. Viterbo, "The land surface-atmosphere interaction: a review based on observational and global modeling perspectives," Journal of Geophysical Research D, vol. 101, no. 3, pp. 7209–7225, 1996.
9. R. A. Pielke Sr. and D. Niyogi, "The role of landscape processes within the climate system," Lecture Notes in Earth Sciences, vol. 115, pp. 67–85, 2010.
10. O. L. Phillips, L. E. Aragão, S. L. Lewis, et al., "Drought sensitivity of the Amazon rainforest," Science, vol. 323, no. 5919, pp. 1344–1347, 2009.
11. Z. M. Subin, W. J. Riley, J. Jin, D. S. Christianson, M. S. Torn, and L. M. Kueppers, "Ecosystem feedbacks to climate change in California: development, testing, and analysis using a coupled regional atmosphere and land surface model (WRF3-CLM3.5)," Earth Interactions, vol. 15, no. 15, pp. 1–38, 2011.
12. M. Wang, X. Zhang, and X. Yan, "Modeling the climatic effects of urbanization in the Beijing-Tianjin-Hebei metropolitan area," in Theoretical and Applied Climatology, Springer, 2012.
13. R. A. Pielke Sr., G. Marland, R. A. Betts et al., "The influence of land-use change and landscape dynamics on the climate system: relevance to climate-change policy beyond the radiative effect of greenhouse gases," Philosophical Transactions of the Royal Society A, vol. 360, no. 1797, pp. 1705–1719, 2002.
14. R. A. Pielke Sr., J. Adegoke, A. Beltrán-Przekurat et al., "An overview of regional land-use and land-cover impacts on rainfall," Tellus B, vol. 59, no. 3, pp. 587–601, 2007.
15. V. K. Arora and A. Montenegro, "Small temperature benefits provided by realistic afforestation efforts,"Nature Geoscience, vol. 4, no. 8, pp. 514–518, 2011.
16. A. M. Degu, F. Hossain, D. Niyogi et al., "The influence of large dams on surrounding climate and precipitation patterns," Geophysical Research Letters, vol. 38, no. 4, Article ID L04405, 7 pages, 2011.

17. M. Wang, X. Yan, J. Liu, and X. Zhang, "The contribution of urbanization to recent extreme heat events and a potential mitigation strategy in the Beijing-Tianjin-Hebei metropolitan area," in Theoretical and Applied Climatology, 2013.
18. A. T. Woldemichael, F. Hossain, R. Pielke, and A. Beltrán-Przekurat, "Understanding the impact of dam-triggered land use/land cover change on the modification of extreme precipitation," Water Resources Research, vol. 48, no. 9, pp. 1–16, 2012.
19. P. Meiyappan and A. K. Jain, "Three distinct global estimates of historical land-cover change and land-use conversions for over 200 years," Frontiers of Earth Science, vol. 6, no. 2, pp. 122–139, 2012.
20. J. Pongratz, C. H. Reick, T. Raddatz, and M. Claussen, "Biogeophysical versus biogeochemical climate response to historical anthropogenic land cover change," Geophysical Research Letters, vol. 37, no. 8, Article ID L08702, 2010.
21. P. Friedlingstein and I. C. Prentice, "Carbon-climate feedbacks: a review of model and observation based estimates," Current Opinion in Environmental Sustainability, vol. 2, no. 4, pp. 251–257, 2010.
22. P. A. Dirmeyer and J. Shukla, "Albedo as a modulator of climate response to tropical deforestation,"Journal of Geophysical Research, vol. 99, no. 10, pp. 20–877, 1994.
23. M. Georgescu, D. B. Lobell, and C. B. Field, "Direct climate effects of perennial bioenergy crops in the United States," Proceedings of the National Academy of Sciences of the United States of America, vol. 108, no. 11, pp. 4307–4312, 2011.
24. J. Hansen, L. Nazarenko, R. Ruedy et al., "Climate Change: earth's energy imbalance: confirmation and implications," Science, vol. 308, no. 5727, pp. 1431–1435, 2005.
25. S. Bathiany, M. Claussen, V. Brovkin, T. Raddatz, and V. Gayler, "Combined biogeophysical and biogeochemical effects of large-scale forest cover changes in the MPI earth system model,"Biogeosciences Discussions, vol. 7, no. 1, p. 387, 2010.
26. J. Jin, S. Lu, S. Li, and N. L. Miller, "Impact of land use change on the local climate over the tibetan plateau," Advances in Meteorology, vol. 2010, Article ID 837480, 6 pages, 2010.

27. F. H. Lambert, M. J. Webb, and M. M. Joshi, "The relationship between land-ocean surface temperature contrast and radiative forcing," Journal of Climate, vol. 24, no. 13, pp. 3239–3256, 2011.
28. V. Brovkin, A. Ganopolski, M. Claussen, C. Kubatzki, and V. Petoukhov, "Modelling climate response to historical land cover change," Global Ecology and Biogeography, vol. 8, no. 6, pp. 509–517, 1999.
29. J. Charney, P. H. Stone, and W. J. Quirk, "Drought in the Sahara: a biogeophysical feedback mechanism," Science, vol. 187, no. 4175, pp. 434–435, 1975.
30. J. Shukla and Y. Mintz, "Influence of land-surface evapotranspiration on the earth's climate," Science, vol. 215, no. 4539, pp. 1498–1501, 1982.
31. E. F. Lambin and H. Geist, Land-Use and Land-Cover Change: Local Processes and Global Impacts, Springer, 2006.
32. V. Brovkin, L. Boysen, T. Raddatz, V. Gayler, A. Loew, and M. Claussen, "Evaluation of vegetation cover and land-surface albedo in MPI-ESM CMIP5 simulations," Journal of Advances in Modeling Earth Systems, pp. 48–57, 2013.
33. J. D. Wickham, T. G. Wade, and K. H. Riitters, "Comparison of cropland and forest surface temperatures across the conterminous United States," Agricultural and Forest Meteorology, vol. 166-167, pp. 137–143, 2012.
34. J. D. Wickham, T. G. Wade, and K. H. Riitters, "Empirical analysis of the influence of forest extent on annual and seasonal surface temperatures for the continental United States," Global Ecology and Biogeography, vol. 22, no. 5, pp. 620–629, 2013.
35. J. Shukla, C. Nobre, and P. Sellers, "Amazon deforestation and climate change," Science, vol. 247, no. 4948, pp. 1322–1325, 1990.
36. Y. Malhi, J. T. Roberts, R. A. Betts, T. J. Killeen, W. Li, and C. A. Nobre, "Climate change, deforestation, and the fate of the Amazon," Science, vol. 319, no. 5860, pp. 169–172, 2008.
37. L. Salvati and S. Bajocco, "Land sensitivity to desertification across Italy: past, present, and future,"Applied Geography, vol. 31, no. 1, pp. 223–231, 2011.

38. S. E. Nicholson, C. J. Tucker, and M. B. Ba, "Desertification, drought, and surface vegetation: an example from the west African sahel," Bulletin of the American Meteorological Society, vol. 79, no. 5, pp. 815–829, 1998.
39. M. Zhao, A. J. Pitman, and T. Chase, "The impact of land cover change on the atmospheric circulation,"Climate Dynamics, vol. 17, no. 5-6, pp. 467–477, 2001.
40. M. Claussen, V. Brovkin, and A. Ganopolski, "Biophysical versus biogeochemical feedbacks of large-scale land cover change," Geophysical Research Letters, vol. 28, no. 6, pp. 1011–1014, 2001.
41. E. L. Davin and N. de Noblet-Ducoudre, "Climatic impact of global-scale Deforestation: radiative versus nonradiative processes," Journal of Climate, vol. 23, no. 1, pp. 97–112, 2010.
42. R. S. Defries, L. Bounoua, and G. J. Collatz, "Human modification of the landscape and surface climate in the next fifty years," Global Change Biology, vol. 8, no. 5, pp. 438–458, 2002.
43. A. E. Dessler, "A determination of the cloud feedback from climate variations over the past decade,"Science, vol. 330, no. 6010, pp. 1523–1527, 2010.
44. A. H. Lynch, F. S. Chapin III, L. D. Hinzman et al., "Surface energy balance on the arctic tundra: measurements and models," Journal of Climate, vol. 12, no. 8, pp. 2585–2606, 1999.
45. F.-S. Chapin III, M. Sturm, M. C. Serreze et al., "Role of land-surface changes in arctic summer warming," Science, vol. 310, no. 5748, pp. 657–660, 2005.
46. C. J. Houldcroft, W. M. F. Grey, M. Barnsley, C. M. Taylor, S. O. Los, and P. R. J. North, "New vegetation Albedo parameters and global fields of soil background Albedo derived from MODIS for use in a climate model," Journal of Hydrometeorology, vol. 10, no. 1, pp. 183–198, 2009.
47. G. B. Bonan and D. Pollard, "vegetation on global climate," Nature, vol. 359, 22 pages, 1992.
48. J. H. Copeland, R. A. Pielke, and T. G. F. Kittel, "Potential climatic impacts of vegetation change: a regional modeling study," Journal of Geophysical Research D, vol. 101, no. 3, pp. 7409–7418, 1996.

49. Y. Sud, J. Shukla, and Y. Mintz, "Influence of land surface roughness on atmospheric circulation and precipitation-a sensitivity study with a general circulation model," Journal of Applied Meteorology, vol. 27, no. 9, pp. 1036–1054, 1988.
50. J. I. House, I. C. Prentice, and C. C. Le Quéré, "Maximum impacts of future reforestation or deforestation on atmospheric CO_2," Global Change Biology, vol. 8, no. 11, pp. 1047–1052, 2002.
51. R. A. Houghton and J. L. Hackler, "Sources and sinks of carbon from land-use change in China," Global Biogeochemical Cycles, vol. 17, no. 2, 1034 pages, 2003.
52. J. Fang, A. Chen, C. Peng, S. Zhao, and L. Ci, "Changes in forest biomass carbon storage in China between 1949 and 1998," Science, vol. 292, no. 5525, pp. 2320–2322, 2001.
53. S. W. Pacala, G. C. Hurtt, D. Baker et al., "Consistent land- and atmosphere-based U.S. Carbon sink estimates," Science, vol. 292, no. 5525, pp. 2316–2320, 2001.
54. G. C. Hurtt, S. W. Pacala, P. R. Moorcroft et al., "Projecting the future of the U.S. Carbon sink,"Proceedings of the National Academy of Sciences of the United States of America, vol. 99, no. 3, pp. 1389–1394, 2002.
55. M. Heimann and M. Reichstein, "Terrestrial ecosystem carbon dynamics and climate feedbacks,"Nature, vol. 451, no. 7176, pp. 289–292, 2008.
56. P. Friedlingstein, P. Cox, R. Betts et al., "Climate-carbon cycle feedback analysis: results from the C4MIP model intercomparison," Journal of Climate, vol. 19, no. 14, pp. 3337–3353, 2006.
57. P. Smith, D. S. Powlson, M. J. Glendining, and J. U. Smith, "Potential for carbon sequestration in European soils: preliminary estimates for five scenarios using results from long-term experiments,"Global Change Biology, vol. 3, no. 1, pp. 67–79, 1997.
58. J. Laothawornkitkul, J. E. Taylor, N. D. Paul, and C. N. Hewitt, "Biogenic volatile organic compounds in the Earth system: tansley review," New Phytologist, vol. 183, no. 1, pp. 27–51, 2009.
59. D. V. Spracklen, B. Bonn, and K. S. Carslaw, "Boreal forests, aerosols and the impacts on clouds and climate," Philosophical

Transactions of the Royal Society A, vol. 366, no. 1885, pp. 4613–4626, 2008.

60. A. Kattenberg, F. Giorgi, H. Grassl, et al., "Climate models-projections of future climate," in Climate Change 1995: The Science of Climate Change. Contribution of Working Group I to the Second Assessment Report of the Intergovernmental Panel on Climate Change, pp. 285–357, 1996.
61. R. A. Pielke Sr., R. Avissar, M. Raupach, A. J. Dolman, X. Zeng, and A. S. Denning, "Interactions between the atmosphere and terrestrial ecosystems: influence on weather and climate," Global Change Biology, vol. 4, no. 5, pp. 461–475, 1998.
62. I.-S. Kang, K. Jin, B. Wang et al., "Intercomparison of the climatological variations of Asian summer monsoon precipitation simulated by 10 GCMs," Climate Dynamics, vol. 19, no. 5-6, pp. 383–395, 2002.
63. D. M. Lawrence, P. E. Thornton, K. W. Oleson, and G. B. Bonan, "The partitioning of evapotranspiration into transpiration, soil evaporation, and canopy evaporation in a GCM: impacts on land-atmosphere interaction," Journal of Hydrometeorology, vol. 8, no. 4, pp. 862–880, 2007.
64. C. B. Field, R. B. Jackson, and H. A. Mooney, "Stomatal responses to increased CO_2: implications from the plant to the global scale," Plant, Cell and Environment, vol. 18, no. 10, pp. 1214–1225, 1995.
65. T. N. Chase, R. A. Pielke, T. G. F. Kittel, R. Nemani, and S. W. Running, "Sensitivity of a general circulation model to global changes in leaf area index," Journal of Geophysical Research D, vol. 101, no. 3, pp. 7393–7408, 1996.
66. R. A. Betts, "Biogeophysical impacts of land use on present-day climate: near-surface temperature change and radiative forcing," Atmospheric Science Letters, vol. 2, no. 1–4, pp. 39–51, 2001.
67. H. Salmun and A. Molod, "Progress in modeling the impact of land cover change on the global climate,"Progress in Physical Geography, vol. 30, no. 6, pp. 737–749, 2006.
68. V. Brovkin, M. Claussen, E. Driesschaert et al., "Biogeophysical effects of historical land cover changes simulated by six Earth system models of intermediate complexity," Climate Dynamics, vol. 26, no. 6, pp. 587–600, 2006.

69. H. Von Storch, E. Zorita, and U. Cubasch, "Downscaling of global climate change estimates to regional scales: an application to Iberian rainfall in wintertime," Journal of Climate, vol. 6, no. 6, pp. 1161–1171, 1993.
70. J. Jin, N. L. Miller, and N. Schlegel, "Sensitivity study of four land surface schemes in the WRF model,"Advances in Meteorology, vol. 2010, Article ID 167436, 11 pages, 2010.
71. Y. Wang, L. R. Leung, J. L. McGregor et al., "Regional climate modeling: progress, challenges, and prospects," Journal of the Meteorological Society of Japan, vol. 82, no. 6, pp. 1599–1628, 2004.
72. M. D. Oyama and C. A. Nobre, "Climatic consequences of a large-scale desertification in northeast Brazil: a GCM simulation study," Journal of Climate, vol. 17, no. 16, pp. 3203–3213, 2004.
73. M. T. Coe, M. H. Costa, and B. S. Soares-Filho, "The influence of historical and potential future deforestation on the stream flow of the Amazon Rivervland surface processes and atmospheric feedbacks," Journal of Hydrology, vol. 369, no. 1-2, pp. 165–174, 2009.
74. T. J. Stohlgren, T. N. Chase, R. A. Pielke Sr., T. G. F. Kittel, and J. S. Baron, "Evidence that local land use practices influence regional climate, vegetation, and stream flow patterns in adjacent natural areas,"Global Change Biology, vol. 4, no. 5, pp. 495–504, 1998.
75. N. S. Diffenbaugh, "Influence of modern land cover on the climate of the United States," Climate Dynamics, vol. 33, no. 7-8, pp. 945–958, 2009.
76. K. E. Trenberth, Climate System Modeling, Cambridge University Press, 1992.
77. K. Mcguffie and A. Henderson-Sellers, A Climate Modelling Primer, Wiley, 2005.
78. A. Henderson-Sellers, R. E. Dickinson, T. B. Durbidge, P. J. Kennedy, K. McGuffie, and A. J. Pitman, "Tropical deforestation: modeling local- to regional-scale climate change," Journal of Geophysical Research, vol. 98, no. 4, pp. 7289–7315, 1993.
79. Y. Malhi, L. E. O. C. Aragão, D. Galbraith et al., "Exploring the likelihood and mechanism of a climate-change-induced dieback of the Amazon rainforest," Proceedings of the National Academy

of Sciences of the United States of America, vol. 106, no. 49, pp. 20610–20615, 2009.

80. Y. Xue and J. Shukla, "The influence of land surface properties on Sahel climate. Part I: desertification,"Journal of Climate, vol. 6, no. 12, pp. 2232–2245, 1993.
81. S. Nicholson, "Land surface processes and Sahel climate," Reviews of Geophysics, vol. 38, no. 1, pp. 117–139, 2000.
82. P. J. Sellers, D. A. Randall, G. J. Collatz et al., "A revised land surface parameterization (SiB2) for atmospheric GCMs. Part I: model formulation," Journal of Climate, vol. 9, no. 4, pp. 676–705, 1996.
83. G. B. Bonan, K. W. Oleson, M. Vertenstein et al., "The land surface climatology of the community land model coupled to the NCAR community climate model," Journal of Climate, vol. 15, no. 22, pp. 3123–3149, 2002.
84. D. Niyogi, K. Alapaty, S. Raman, and F. Chen, "Development and evaluation of a coupled photosynthesis-based gas exchange evapotranspiration model (GEM) for mesoscale weather forecasting applications," Journal of Applied Meteorology and Climatology, vol. 48, no. 2, pp. 349–368, 2009.
85. Y. Xue, F. J. Zeng, K. E. Mitchell, Z. Janjic, and E. Rogers, "The impact of land surface processes on simulations of the U.S. hydrological cycle: a case study of the 1993 flood usingthe SSiB land surface model in the NCEP Eta regional model," Monthly Weather Review, vol. 129, no. 12, pp. 2833–2860, 2001.
86. Z. Yiqun, Q. Yongfu, M. Manqian, Y. Ge, K. Yushou, and Z. Donghua, "The effects of vegetation change on regional climate I: simulation results," Acta Meteorologica Sinica, vol. 60, no. 1, pp. 1–16, 2002.
87. Y. Dai, X. Zeng, R. E. Dickinson et al., "The common land model," Bulletin of the American Meteorological Society, vol. 84, no. 8, pp. 1013–1023, 2003.
88. K. Klein Goldewijk, A. Beusen, G. Van Drecht, and M. De Vos, "The HYDE 3.1 spatially explicit database of human-induced global land-use change over the past 12,000 years," Global Ecology and Biogeography, vol. 20, no. 1, pp. 73–86, 2011.

89. P. Kabat, Vegetation, Water, Humans and the Climate: A New Perspective on an Interactive System, Springer, 2004.

90. L. M. Kueppers, M. A. Snyder, and L. C. Sloan, "Irrigation cooling effect: regional climate forcing by land-use change," Geophysical Research Letters, vol. 34, no. 3, Article ID L03703, 5 pages, 2007.

91. E. M. Douglas, A. Beltrán-Przekurat, D. Niyogi, R. A. Pielke Sr., and C. J. Vörösmarty, "The impact of agricultural intensification and irrigation on land-atmosphere interactions and Indian monsoon precipitation—a mesoscale modeling perspective," Global and Planetary Change, vol. 67, no. 1-2, pp. 117–128, 2009.

92. L. M. Kueppers, M. A. Snyder, L. C. Sloan et al., "Seasonal temperature responses to land-use change in the western United States," Global and Planetary Change, vol. 60, no. 3-4, pp. 250–264, 2008.

93. D. B. Lobell and C. Bonfils, "The effect of irrigation on regional temperatures: a spatial and temporal analysis of trends in California, 1934–2002," Journal of Climate, vol. 21, no. 10, pp. 2063–2071, 2008.

94. E. M. Douglas, D. Niyogi, S. Frolking et al., "Changes in moisture and energy fluxes due to agricultural land use and irrigation in the Indian Monsoon Belt," Geophysical Research Letters, vol. 33, no. 14, Article ID L14403, 5 pages, 2006.

95. D. Lobell, G. Bala, A. Mirin, T. Phillips, R. Maxwell, and D. Rotman, "Regional differences in the influence of irrigation on climate," Journal of Climate, vol. 22, no. 8, pp. 2248–2255, 2009.

96. J. M. Shepherd, "A review of current investigations of urban-induced rainfall and recommendations for the future," Earth Interactions, vol. 9, no. 12, pp. 1–27, 2005.

97. B. Stone Jr., "Land use as climate change mitigation," Environmental Science and Technology, vol. 43, no. 24, pp. 9052–9056, 2009.

98. I. D. Stewart and T. R. Oke, "Local climate zones for urban temperature studies," Bulletin of the American Meteorological Society, vol. 93, no. 12, pp. 1879–1900, 2012.

99. J. Li, C. Song, L. Cao, F. Zhu, X. Meng, and J. Wu, "Impacts of landscape structure on surface urban heat islands: a case study of

Shanghai, China," Remote Sensing of Environment, vol. 115, no. 12, pp. 3249–3263, 2011.

100. S. Kato and Y. Yamaguchi, "Estimation of storage heat flux in an urban area using ASTER data," Remote Sensing of Environment, vol. 110, no. 1, pp. 1–17, 2007.

101. J. K. W. Wong and L. S. -K. Lau, "From the "urban heat island" to the vgreen island"? A preliminary investigation into the potential of retrofitting green roofs in Mongkok district of Hong Kong," Habitat International, vol. 39, pp. 25–35, 2013.

102. K. W. Oleson, G. B. Bonan, J. Feddema, and M. Vertenstein, "An urban parameterization for a global climate model. Part II: sensitivity to input parameters and the simulated urban heat island in offline simulations," Journal of Applied Meteorology and Climatology, vol. 47, no. 4, pp. 1061–1076, 2008.

103. A. Murata, H. Sasaki, M. Hanafusa, and K. Kurihara, "Estimation of urban heat island intensity using biases in surface air temperature simulated by a nonhydrostatic regional climate model," Theoretical and Applied Climatology, vol. 112, no. 1-2, pp. 351–361, 2013.

104. K. Wang, H. Ye, F. Chen, Y. Xiong, and C. Wang, "Urbanization effect on the diurnal temperature range: different roles under solar dimming and brightening," Journal of Climate, vol. 25, no. 3, pp. 1022–1027, 2012.

105. A. M. Coutts, J. Beringer, and N. J. Tapper, "Impact of increasing urban density on local climate: spatial and temporal variations in the surface energy balance in Melbourne, Australia," Journal of Applied Meteorology and Climatology, vol. 46, no. 4, pp. 477–493, 2007.

106. Y. Ezber, O. L. Sen, T. Kindap, and M. Karaca, "Climatic effects of urbanization in Istanbul: a statistical and modeling analysis," International Journal of Climatology, vol. 27, no. 5, pp. 667–679, 2007.

107. K. Trusilova, M. Jung, G. Churkina, U. Karsten, M. Heimann, and M. Claussen, "Urbanization impacts on the climate in Europe: numerical experiments by the PSU-NCAR mesoscale model (MM5)," Journal of Applied Meteorology and Climatology, vol. 47, no. 5, pp. 1442–1455, 2008.

108. S. I. Bohnenstengel, S. Evans, P. A. Clark, and S. E. Belcher, "Simulations of the London urban heat island," Quarterly Journal of the Royal Meteorological Society, vol. 137, no. 659, pp. 1625–1640, 2011.

109. D.-L. Zhang, Y.-X. Shou, and R. R. Dickerson, "Upstream urbanization exacerbates urban heat island effects," Geophysical Research Letters, vol. 36, no. 24, Article ID L24401, 5 pages, 2009.

110. S. Miao, F. Chen, Q. Li, and S. Fan, "Impacts of urban processes and urbanization on summer precipitation: a case study of heavy rainfall in Beijing on 1 August 2006," Journal of Applied Meteorology and Climatology, vol. 50, no. 4, pp. 806–825, 2011.

Chapter 6

Analysis of Climate and Land Use Changes Impacts on Land Degradation in the North China Plain

Zhihui Li[1, 2, 3], Xiangzheng Deng[1, 2], Fang Yin[4], and Cuiyuan Yang[5]

[1]Institute of Geographic Sciences and Natural Resources Research, Chinese Academy of Sciences, Beijing 100101, China

[2]Center for Chinese Agricultural Policy, Chinese Academy of Sciences, Beijing 100101, China

[3]University of Chinese Academy of Sciences, Beijing 100049, China

[4]Leibniz Institute of Agricultural Development in Transition Economies (IAMO), Theodor-Lieser Street 2, 06120 Halle (Saale), Germany

[5]College of Geomatics, Shandong University of Science and Technology, Qingdao, Shandong 266590, China

ABSTRACT

Land degradation is a complex process which involves both the natural ecosystem and the socioeconomic system, among which climate and land use changes are the two predominant driving factors. To comprehensively and quantitatively analyze the land degradation process, this paper employed the Normalized Difference Vegetation Index (NDVI) as a proxy to assess land degradation and further applied the binary panel logit regression model to analyze the impacts of the driving factors on land degradation in the North China Plain. The results revealed that an increase in rainfall and temperature would significantly and positively contribute to the land improvement, and conversion from cultivated land to grassland and forest land showed positive relationship with land improvement, while conversion to built-up area will lead to land degradation. Besides, human agricultural intensification represented by fertilizer utilization will help to improve the land quality. The economic development may exert positive impacts on land quality to alleviate land degradation, although the rural economic development and agricultural production will exert negative impacts on the land and lead to land degradation. Infrastructure construction would modify the land surface and further resulted in land degradation. The findings of the research will provide scientific information for sustainable land management.

INTRODUCTION

Land degradation has become a critical issue worldwide, especially in the developing countries, which leads to great concerns about food security. To improve livelihoods of human beings and to keep sustainable development of human society, healthy land ecosystems are basically essential elements. However, the key services of good quality land and their true values have been usually taken for granted and underestimated, leading to serious land degradation, which not only deteriorates the ecosystem services but also hinders regional sustainable development. Land degradation means a significant reduction of the productive capacity of land. It is an interactive process involving various casual factors, among which climate changes, land use/cover changes, and human dominated land management play a significant role [1–4]. Land degradation involves two interlocking complex systems, the natural ecosystem and the human social system, and both changes in biophysical natural ecosystem and socioeconomic conditions will affect the land degradation process [5]. Natural forces affect land degradation through periodic stresses of extreme and persistent climatic events. Human activities also contribute to land degradation through deforestation, removal of natural vegetation, and urban sprawl that lead to land use/cover changes; unsustainable agricultural land use management practices, such as use and abuse of fertilizer, pesticide, and heavy machinery; and overgrazing, improper crop rotation, poor irrigation practices, and so forth [6].

In comparison to other regions, land degradation afflicts China more seriously in terms of the extent, intensity, economic impacts, and affected population [7]. China accounts for 22% of the world's population but only 6.4% of the global land area and 7.2% of the global cultivated land area. Agricultural production is an important issue that always plagues the national economy and livelihood of citizens [8]. However, the rapid population growth and urbanization, unreasonable human utilization, and influence of natural factors have caused degradation of 5.392 million km^2 of land, which accounts for about 56.2% of the total national area [9]. Only 1.3 million km^2 of land is suitable for cultivation in China, which accounts for about 14% of the total national land area. The land degradation has involved more than half of the total cultivated land area and further exerts more pressure on the economic benefits of agricultural production and food security. Besides, as the cultivated land degradation will directly affect the potential land productivity, more inputs such as fertilizer and irrigation water will be needed in order to get the same production as well as yield level, which will increase the production cost [10]. In addition, the structural change and pattern succession of the land system resulting from land conversions will undoubtedly lead to changes in the suitability and quality of land and directly influence land productivity [11]. Hence sustainable productive land management is crucial for the country's long-term agricultural economy.

China is a large country with significant spatial variation of natural/climatic conditions and diverse socioeconomic characteristics. For example, the eastern, central, and western part of China have different population density, industrial structure, per capita income, and so forth. The difference among regions will affect the land use, leading to difference in the way and extent of economic use of land resources [12]. The North China Plain has been one of the most productive agricultural bases in China for thousands of years. It is of great significance to quantitatively measure the intensity and spatial pattern of land degradation and analyze the driving mechanisms, which can provide information for sustainable land management to ensure the food safety in the North China Plain.

In order to provide information for sustainable land management, it is necessary to quantitatively clarify the impacts of major natural and socioeconomic driving forces on land degradation. The recent development of remote sensing techniques (RS) and geographic information system (GIS) techniques has enhanced the capabilities to obtain and handle spatial information on the heterogeneities of land surface characteristics. The integration of RS and GIS technologies has been proven to be an efficient approach and has been successfully used in various investigations for land degradation assessment. Land degradation is a long-term process indicating the loss of ecosystem function and productivity. For quantitative analysis of land degradation, satellite measurements of the Normalized Difference Vegetation Index (NDVI), which is the most widely used proxy for vegetation cover and production, have been widely applied in studies of land degradation from the field scale to national and global scales [7, 13–15]. Therefore, the objective of this study is to explore the spatial pattern and temporal variation of land degradation based on NDVI in the North China Plain and then to analyze the impact mechanisms of a set of natural and socioeconomic driving forces on land degradation, with the application of the binary panel logit regression model.

DATA PROCESSING AND ANALYSES

An Overview of the North China Plain

The North China Plain (32°08′–40°16′N, 112°10′–122°40′E) is one of the most significant grain production regions of China. It covers two metropolises (Beijing and Tianjin) and five provinces including Anhui, Hebei, Henan, Jiangsu, and Shandong. As it is an alluvial plain developed by the intermittent flooding of the Huanghe, Huaihe, and

Haihe rivers, it is also known as Huang-Huai-Hai Plain in China (Figure 1).

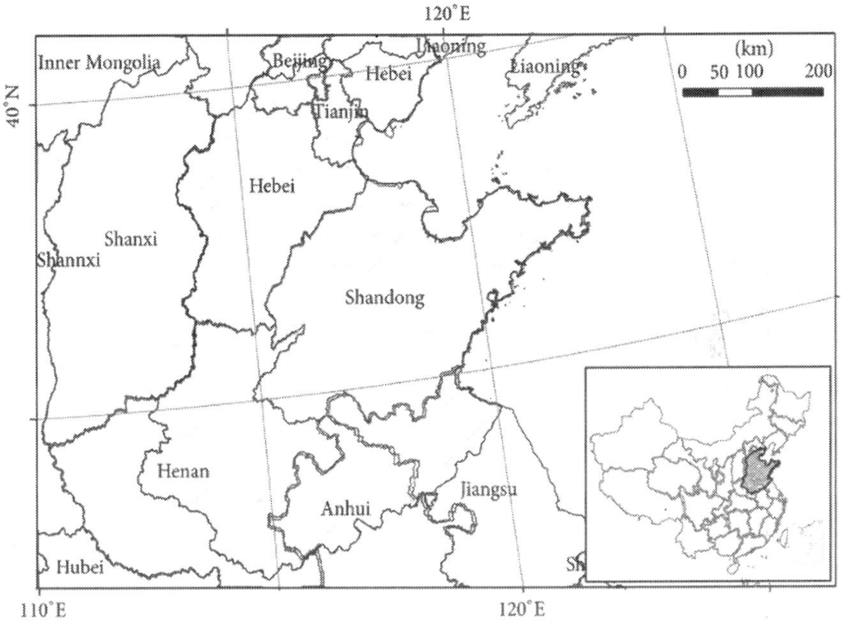

Figure 1: Location of the North China Plain.

As estimated with the remote sensing data derived from the Landsat TM and ETM+ images in the year 2008, the cultivated land accounts for about 68% of the total land area of the North China Plain (Figure 2). There is a semiarid to semihumid warm temperate climate in the North China Plain, with the average annual temperature of 10°C–15°C and annual rainfall of 500–800 mm. It is evident that the North China Plain is suitable for developing rain-fed agriculture and planting a variety of crops and fruit trees. Meanwhile, it is conducive to mechanization and irrigation with its natural conditions of a flat terrain, deep soil, and contiguous land concentration. All of these unique conditions have made the North China Plain a major agricultural region and a significant commodity base for grain, cotton, meat, and oil in China [16].

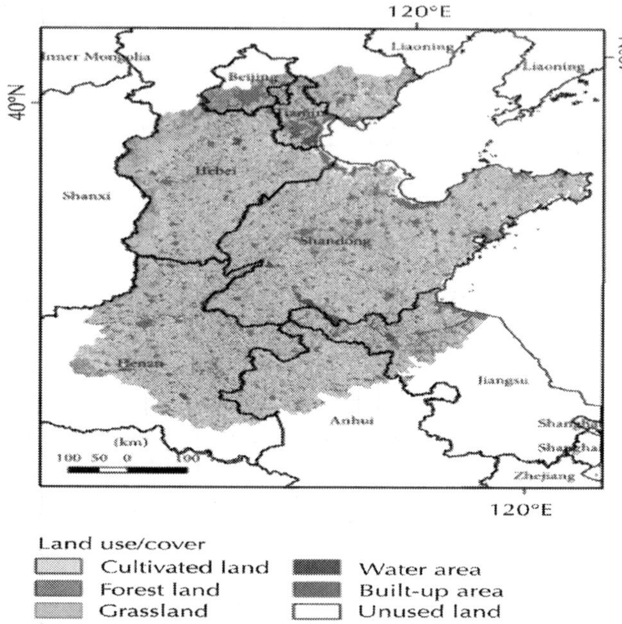

Figure 2: Land use pattern of the North China Plain in the year 2008.

However, the rapid population growth and urban land expansion have led to serious cultivated land loss and environmental degradation in the North China Plain. Many researches showed that the conversion from cultivated land to other types of land led to the decrease of cultivated land in the North China Plain [11, 16,17]. Among the researches, Shi et al. (2013) pointed out that the area of cropland abandonment was much greater than the area of cropland reclamation in the North China Plain [17]. Besides, the quality of cultivated land converted to other land use types is generally higher than that of the cultivated land converted from other land use types [11]. As a result, the shrinkage of cultivated land in the North China Plain will exert great impact on the land quality. How to improve land productivity is becoming vital for the improvement of agricultural production. In addition, though the statistic data showed that the grain yields were increasing, it is due to the additional investment into the agricultural production. In this regard, it is an urgent issue to study how to make scientific policy and sustainable land management measures to rationally regulate the land conversion and control land degradation to improve the productivity of the remaining resources in the agricultural sector.

Data Preparation

According to previous studies, driving factors of land degradation are complex, among which biophysical causes (including topography that determines soil erosion hazard and climatic conditions, such as rainfall and temperature) and land use change (such as deforestation, desertification resulting from unsustainable land management practices) are major direct contributors to land degradation, while the underlying causes of land degradation include population density, economic development, land tenure, and access to agricultural extension, and infrastructure. Thus, in order to analyze the situation of agricultural production and impact mechanisms of climate and land use changes integrated with socioeconomic factors on cultivated land degradation in the North China Plain, we collected the NDVI data, land use datasets, climatic data, socioeconomic statistical data, and other geophysical data of the years 2000–2008.

NDVI

NDVI is one of the most commonly used indicators for vegetation monitoring [18, 19]. Many previous studies have used the NDVI and other measures based on the NDVI as indicators of changes in ecosystem productivity and land degradation, which showed that NDVI is strongly correlated with the green vegetation coverage [20]. Though there are some limitations and criticisms related to the use of this measure, NDVI remains the only dataset available at the global level and the only dataset that reliably provides information about the condition of the aboveground biomass [21, 22]. In this study, we used the remote sensing data of NDVI to measure the land degradation. Before 2006, the NDVI data was obtained from the GIMMS NDVI dataset produced by the GLCF (Global Land Cover Facility) research group at the University of Maryland from July 1981 to December 2006, with a spatial resolution of 8 km by 8 km and a temporal resolution of 15 days. The NDVI data after the year 2006 was derived from SPOT VEGETATION data (ten-day ensembles) which were derived from a dataset developed over the period from April 1998 to December 2010, with a spatial resolution of l km by l km. Based on the period when the two datasets overlapped (1998 to 2006), we conducted data processing according to the previous experience of Yin et al. (2014) [23]. Firstly, correlation analysis between the maximum monthly NDVI and other factors was performed, following which a linear regression equation was established to extend the GIMMS (Global Inventory Modeling and Mapping Studies) dataset from 2007 to 2008. This helps to eliminate any sensor errors in the other two datasets [24]. To ensure the data quality, reliable and internationally recognized data pretreatment processes were used [25]. In order to eliminate cloud contamination effects and the noise caused by other atmospheric phenomena, we included a smoothing method proposed by Chen et al. [26] based on the Savitzky-Golay filter. Finally, we clipped the NDVI data of the North China Plain during the period from the year 2000 to 2008.

Land Use

The land use data came from the land use database of the Resources and Environment Scientific Data Center, Chinese Academy of Sciences [11, 27, 28]. The database was constructed from remotely sensed digital images by the US Landsat TM/ETM satellite with a spatial resolution of 30 by 30 meters. Further the land use data are upscaled to 1 by 1 kilometer. The data that was analyzed in this study covered the years 2000 and 2008.

Climatic Data

The climatic data, mainly including temperature and rainfall from the year 2000 to 2008, were derived from the daily records of 117 observation stations covering the entire area of the North China Plain, which were maintained by the China Meteorological Administration. A spline interpolation using the coupling-fitted thin plate interpolation method was chosen to interpolate the meteorological data [29]. Based on the climatic dependence on topography, climatic variables were interpolated with spatial resolution of 1 by l kilometer [30]. At a broader scale, temperature declined roughly with the increasing elevation at a typical standard lapse rate of 6.5°C per 1 kilometer [31], so the interpolated air temperature data were adjusted to the sea level according to the altitudinal correction factors based on the digital elevation model (DEM) dataset.

Soil Organic Matter

The soil fertility map of the study area was clipped from the soil map of China provided by the Institute of Soil Science, Chinese Academy of Sciences. It was based on the maps on the scale of 1 : 4,000,000 from the Second National Soil Survey of China for Soil Organic Matter (SOM).

Socioeconomic Statistical Data

In this study, we collected and processed the socioeconomic data of 385 major counties in the North China Plain at a county level for the period of the years 2000–2008, including the gross domestic production (GDP), population, rural farmers' per capita income, fertilizer utilization, pesticide utilization, farm machinery, total grain production, and total grain sown area, which were all acquired from socioeconomic statistical yearbooks published by National Statistical Bureau of China.

Geophysical Data

The geophysical data is also closely related with the land productivity. In this specific study, the geographic factors mainly include elevation and average slope, the data of which were derived from China's Digital Elevation Model dataset supplied by the National Geomatics Center of China. Another part of the geophysical data was the proximity variables, including the distance from each grid to the nearest expressway or highway and other ways, which were incorporated to measure the impacts of the infrastructure construction on the land degradation.

Primary Analysis

Changes of Grain Production

We analyzed the changes of grain production and the relationship between total grain production and grain yield (kg/ha) of the North China Plain during the years 2000–2008. The spatial distribution of grain yield at municipality or county level in the years 2000, 2005, and 2008 is shown in Figure3. The grains mainly included wheat and maize, but some rice, soybean, and tuber crops were also included.

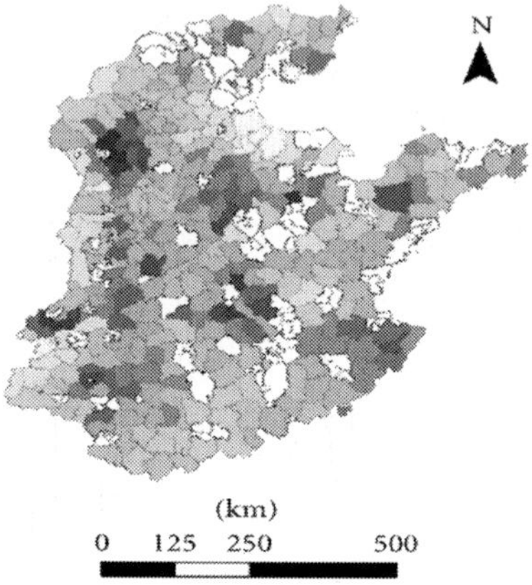

(a)

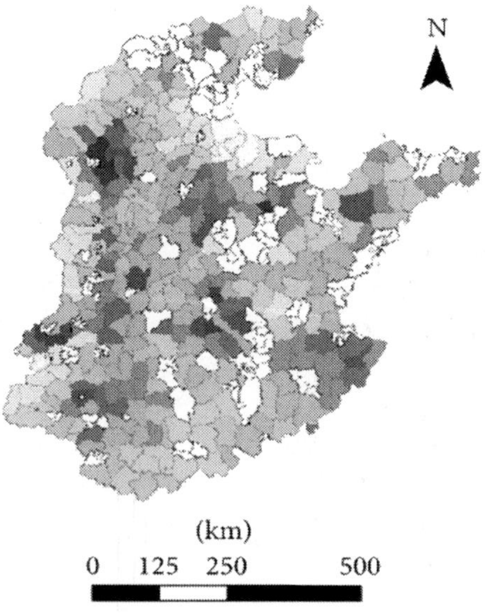

(b)

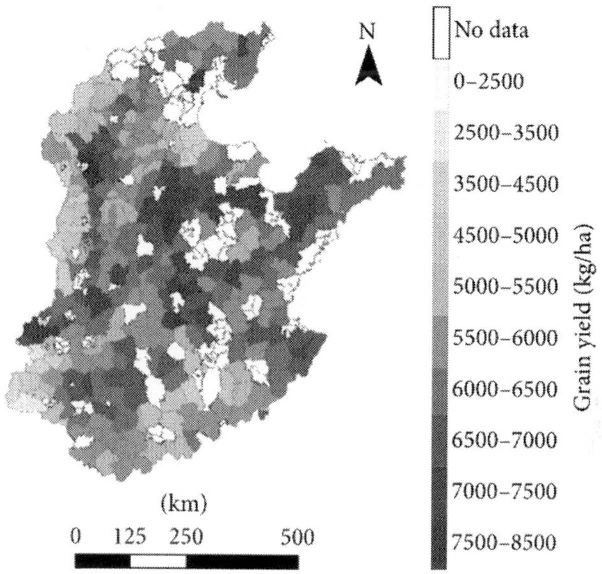

(c)

Figure 3: County-level grain yield of the North China Plain, (a) 2000, (b) 2005, and (c) 2008.

During the period of the years 2000–2008, annual total grain production of the selected counties of the North China Plain increased by more than 40%, and the average grain yield increased from 4994 to 5986 kg/ha (with an increasing rate of 20%). The time-series changing trend of the total grain production is similar to that of the average grain yield, and therefore the increase of agricultural production can be mainly ascribed to the increase in the average grain yield (Figure 4). The time-series change of grain yield from year 2000 to 2008 can be divided into two periods. Firstly, the average grain yield decreased after the year 2000 because of a decrease in grain price. Secondly, the price of grain increased again through governmental support after the year 2003, which made the average yields recover again.

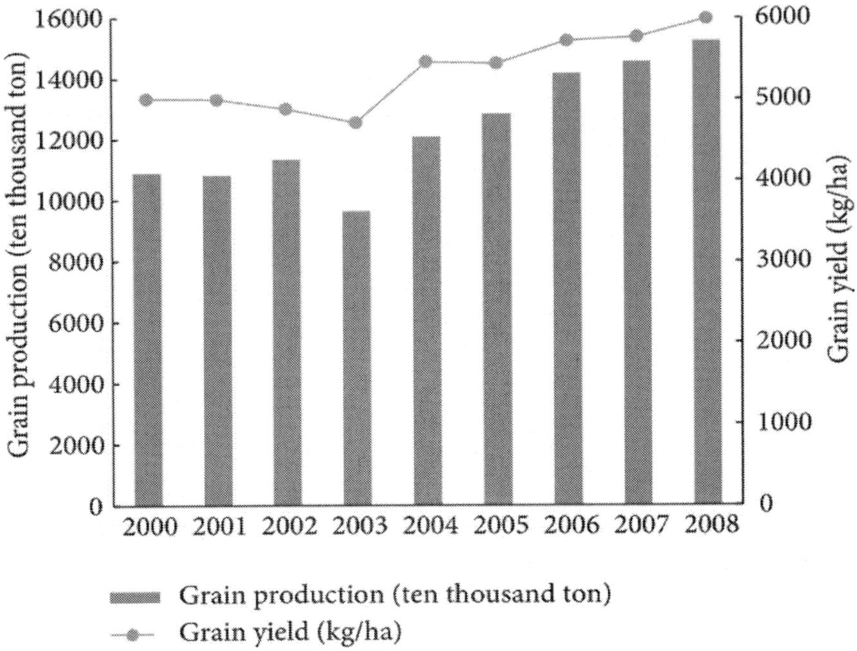

Figure 4: Relationship between grain production and grain yield of the North China Plain, 2000–2008.

Also, we showed that the application rates of fertilizer and pesticide increased with fluctuations from 525 to 622 kg/ha and from 12.6 to 16.2 kg/ha during the period of the years 2000–2008, respectively. The application rate of the farm machinery also showed an increasing trend (Figure 5). The increases of the application of farm machinery, fertilizer, and pesticide can be the key determinants of the increase in crop yield; however, further increases in fertilizer and pesticide application are unlikely to be as effective as they previously were in increasing the grain yields because of the gradually diminishing return rate.

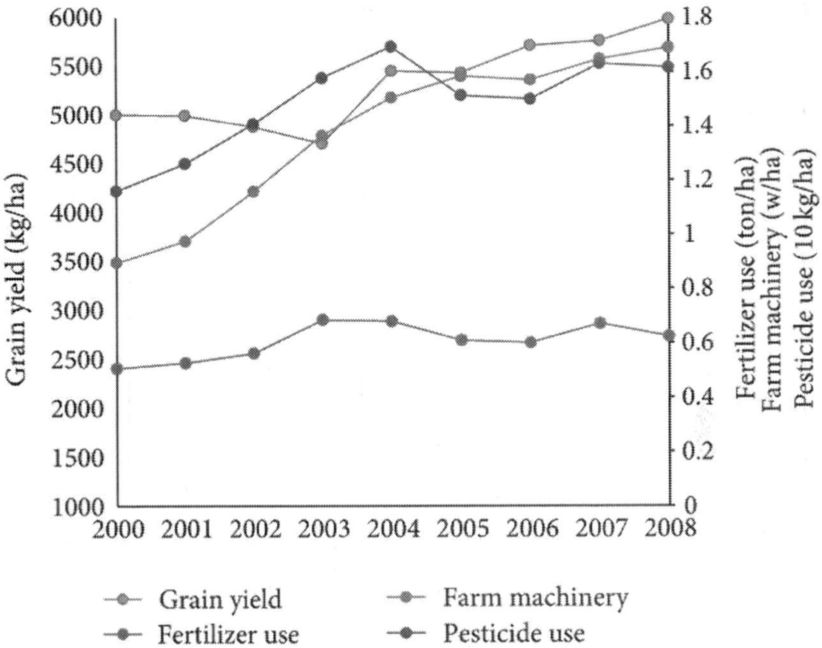

Figure 5: Changes of grain yield, fertilizer use, pesticide use, and farm machinery application of the North China Plain, 2000–2008.

Trends in NDVI

The annual average greenness (represented with NDVI) is chosen as the standard proxy of annual average biomass productivity, which generally fluctuates with the rainfall. The NDVI calculated using remote sensing image data currently is one of the most commonly used indicators to monitor regional or global vegetation and the ecosystem and is the best instruction factor to reflect vegetation growth conditions and coverage [32]. In this study, the cultivated land degradation areas were identified with a sequence of analyses of the NDVI data. The result shows that the annual average NDVI of the North China Plain showed an overall increasing trend during the years 2001–2004. About 90% of the study area's annual average NDVI experienced an increasing trend. While only 5% of the study area's annual average NDVI showed a decreasing trend, which was distributed in the Shandong Province

(Figure 6(a)). In comparison, during the years 2004–2008, about 80% of the study area's annual average NDVI experienced a decreasing trend, with only 19% of the study area's annual average NDVI showed an increasing trend (Figure 6(b)). Besides, the annual average NDVI in some areas showed first increasing and then decreasing trend during the years 2000–2008. Overall, the annual average NDVI showed an increasing trend in most parts of the whole study area during the years 2000–2008 (Figure 6(c)).

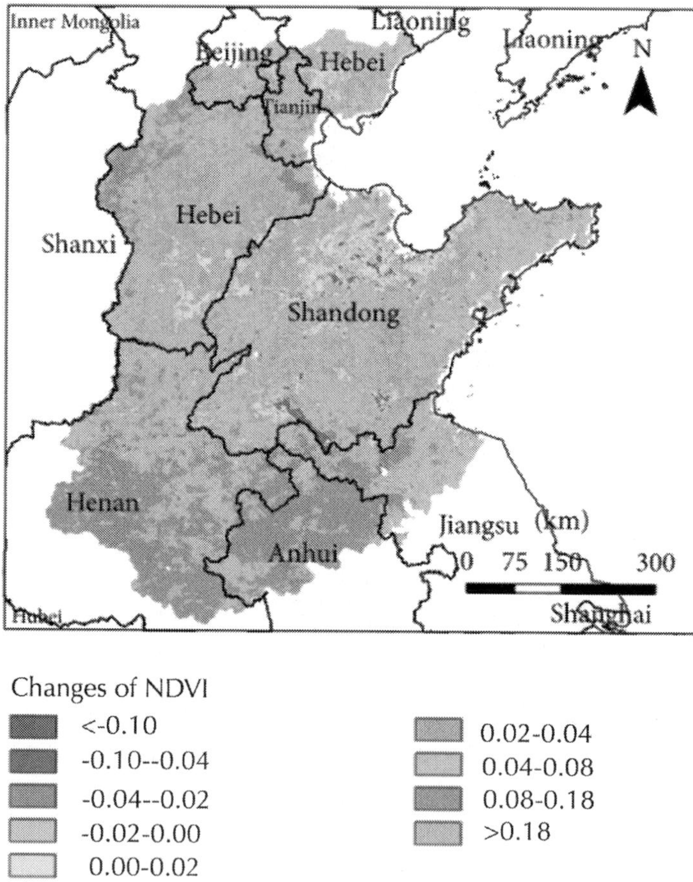

(a)

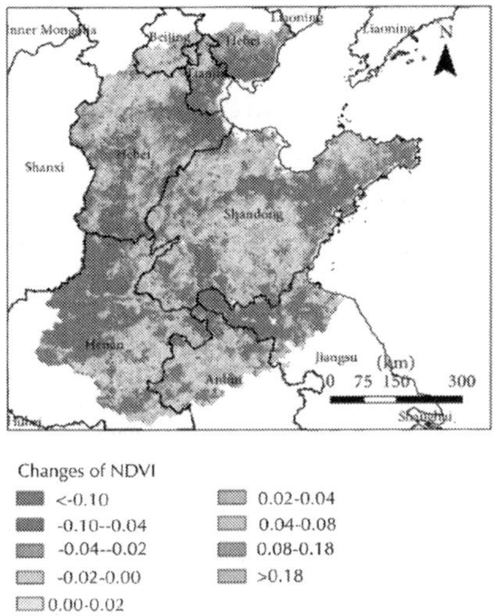

(b)

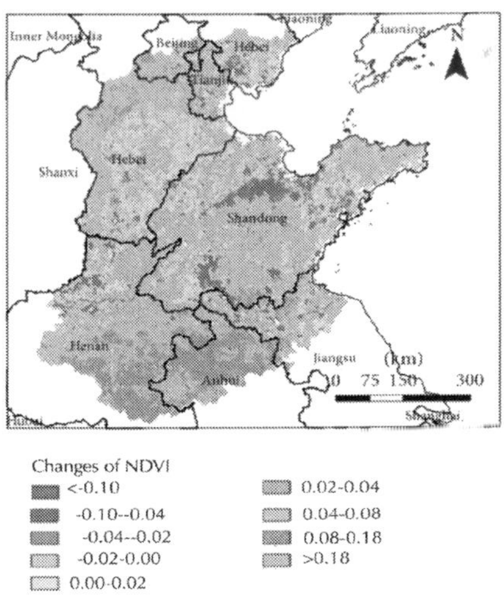

(c)

Figure 6: Changes of NDVI of the North China Plain during (a) 2000–2004; (b) 2004–2008; and (c) 2000–2008.

Land Use Changes

During the years 2000–2008, the land use change is mainly dominated by the conversion from cultivated land to other land use types, especially the conversion from cultivated land to built-up land. As cultivated land is the major land use types in the study area, thus we focus on the cultivated land conversion. In the North China Plain, cultivated land is mainly converted to built-up land (67%), forest land (9.1%), and grassland (12.4%) (Figure 7).

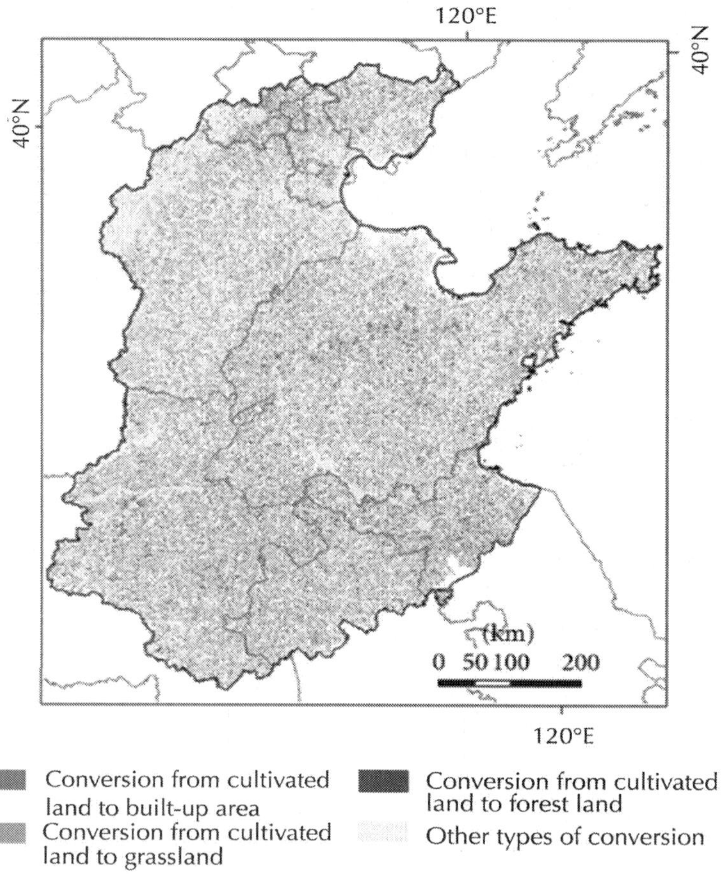

Figure 7: Land conversion from cultivated land to three major land use types during 2000–2008.

Associations between Climate Factors and Land Degradation

We investigated the relationship between NDVI (from year 2000 to 2008) and climate factors. The relationship between NDVI and climate indicators is widely analyzed in many previous studies, which have confirmed that vegetation changes are closely related to changes in the climate factors at the global or regional scale. In particular, it is widely acknowledged that the temperature and rainfall are the most significant factors influencing the ecosystem characteristics and distribution. We sampled the value of the NDVI, rainfall, and temperature and calculated their annual average values in the whole North China Plain. The mean NDVI, temperature, and rainfall for the study area were plotted along a nine-year time series (Figure 8). Figure 8visually showed that the average NDVI experienced a fluctuated increasing trend, and the average rainfall and temperature slightly showed an increasing trend with some fluctuations.

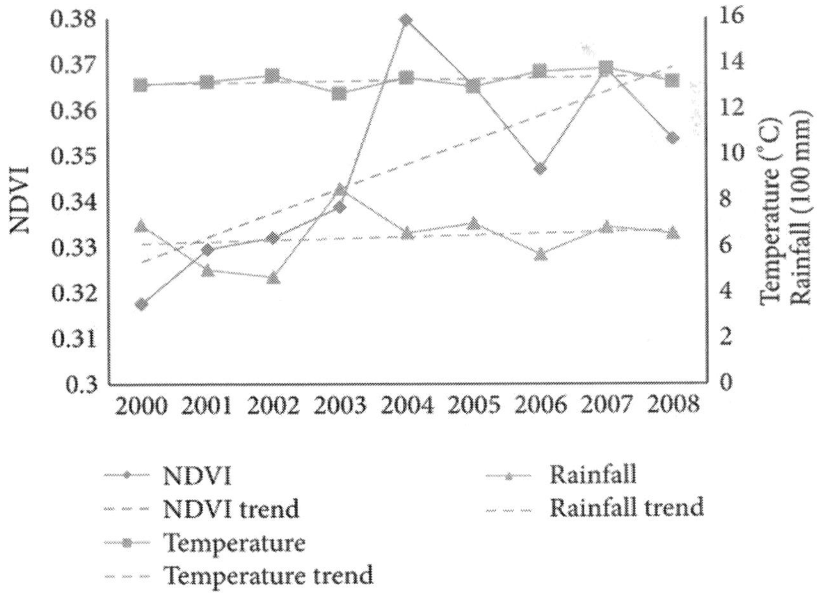

Figure 8: Changing trend of spatially aggregated annual NDVI, rainfall, and temperature.

Specifically, to study the responses of NDVI to the climate change, we analyzed the spatial correlation of NDVI with rainfall and temperature using the correlation analysis at the pixel scale. Figure 9 showed the spatial distribution of the correlation of NDVI with rainfall and temperature, respectively. It shows that the rainfall has a strong effect on NDVI, and there was a positive correlation between rainfall and NDVI in most parts of the study area, especially in the northern part of the study area where it showed a high correlation. As to the relationship between NDVI and temperature, there was generally a positive relationship between NDVI and temperature, which only in the northern part of the study area showed negative relationships.

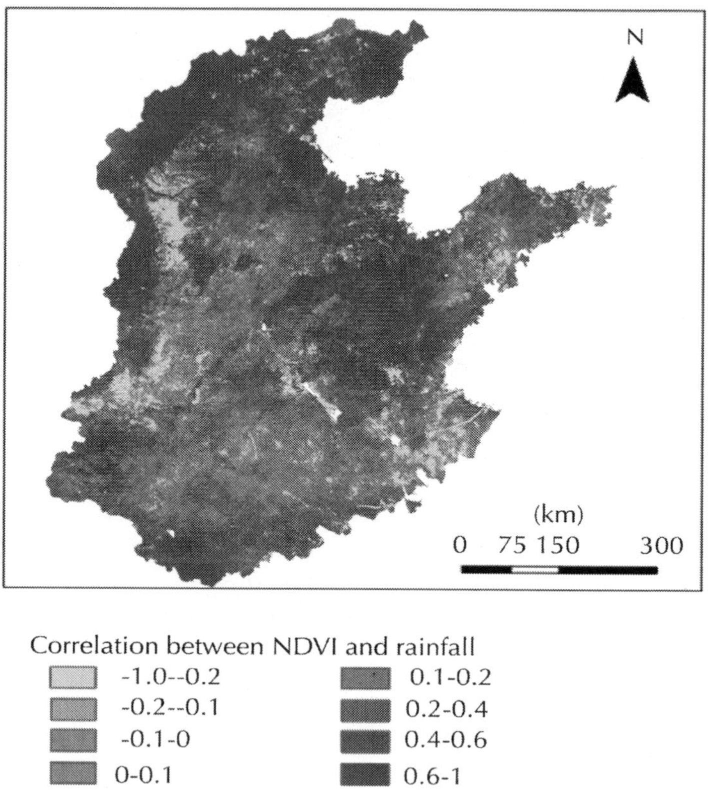

(a)

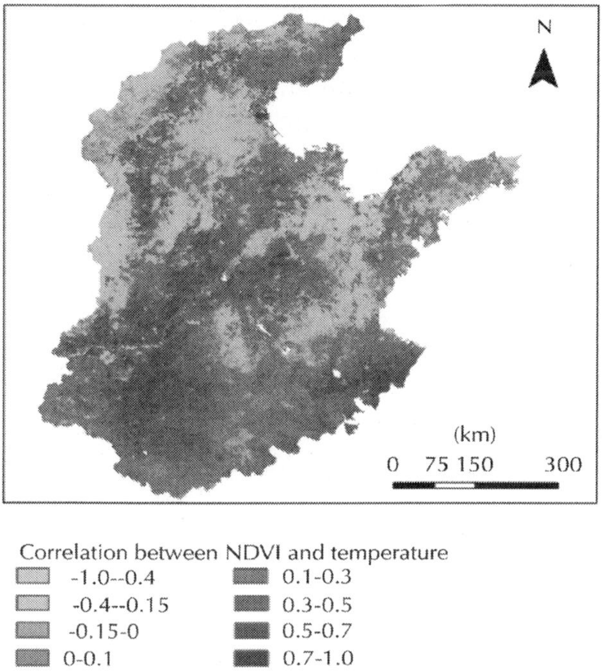

(b)

Figure 9: Spatial patterns of correlation (a) between annual average NDVI and rainfall and (b) between annual average NDVI and temperature at the pixel level in the North China Plain.

ASSESSMENTS OF THE DRIVING FORCES OF LAND DEGRADATION

The causes of land degradation are numerous, interrelated, and complex. The relationship between direct and underlying causes of land degradation is complex also, and the impact of underlying factors is context specific [33]. Quite often, the same cause may lead to diverging consequences in different contexts because of its varying interactions with other proximate and underlying causes of land degradation. This implies that targeting only one underlying factor is not sufficient in itself to address land degradation. It is necessary to take into account a number of underlying and proximate factors when designing policies to prevent or mitigate land degradation. For our model specification, it is essential to identify the effects of various combinations and interactions of underlying and proximate causes of land degradation in a robust manner rather than only analyze the individual cause of land degradation. In this regard, we try to identify the key underlying and proximate causes of land degradation. In this part, we conducted econometric analysis of the causes of land degradation in the North China Plain with the binary logistic regression model.

We constructed the model for estimating causes of land degradation or land improvement at the county level, using annualized data, and the model specification is as follows:

$$\text{Logit}\, Y_t = \beta_0 + \beta_1 \Delta x_{1t} + \beta_2 \Delta x_{2t} + \beta_3 x_3 + \beta_4 x_4 + \varepsilon_{it}, \tag{1}$$

where Y_t=1, if $NDVI_t$-$NDVI_{t-1}$ ≥ 0 and Y_t=0 if;$NDVI_T$-$NDVI_{t-1}$<0; x_{1t}=a vector of biophysical causes of land degradation (e.g., climate conditions, topography, and soil fertility constraints); x_{2t}=a vector of demographic and socioeconomic causes of land degradation (e.g., population density, per capita GDP, rural farmers' per capita income, agricultural intensification); x_3=a vector of variables representing access to infrastructure services (e.g., distance to expressway, distance to high way, and distance to other ways); and x_4=a vector of variables representing land use changes (e.g., county-level percentage of land conversion from cultivated land to built-up land, forest land, and grassland during the years 2000–2008).

Y_t is a binary variable that is being used to identify whether land degradation happens; if Y_t equals or is larger than zero, then we assume that there happens land improvement; otherwise there happens land degradation. Furthermore, we applied a binary panel logit regression model to simultaneously study the relationship between land degradation conditions and all other variables. Table 1 illustrates the results of the regression. The results are to be interpreted with extreme caution due to the complex and multidirectional relationship between NDVI changes and the selected variables. Note that the results might not indicate a causal relationship but only an association between NDVI and the selected biophysical and socioeconomic variables [21].

Table 1: Results for binary panel logistic regression analyses for the biophysical and socioeconomic driving forces of land degradation

Land degradation conditions Y (1 represents land improvement and 0 represents land degradation)	Coef.
Rainfall (mm)	3.221***
Temperature (°C)	1.025***
Percentage of land conversion from cultivated land to built-up land (%)	−0.326
Percentage of land conversion from cultivated land to forest land (%)	1.788
Percentage of land conversion from cultivated land to grassland (%)	8.507*
Fertilizer utilization per unit area (kg/ha)	0.506**
Population density (ten thousand people/km2)	−11.282
Rural farmers' per capita income (ten thousand yuan)	-0.076
Per capita gross domestic production (ten thousand yuan)	0.0004
Share of production value of agriculture animal husbandry and fishery in GDP (%)	−1.672*
DEM (km)	3.214
Slope (degree)	−0.018
Soil organic matter (%)	0.446***
Distance to highway (km)	0.002*
_cons	0.032

Note: *significant at the 10% level; **significant at the 5% level; and ***significant at the 1% level.

The climate changes are significantly related with the land degradation conditions which are represented by the changes in NDVI. As expected, increments in temperature and rainfall are positively related with the increases in NDVI, and increases in temperature and rainfall will significantly contribute to land improvement. Land use change is also the major driving force of land degradation. The results showed that the conversion from cultivated land to grassland and forest land will contribute to land improvement, which corresponded to increase in NDVI, while the conversion from cultivated land to built-up land would lead to land degradation. Among the land conversions, only the conversion from cultivated land to grassland significantly affects the land degradation. In comparison, climate change plays a much more important role than land use change in land degradation. Besides, agricultural intensification represented by fertilizer utilization is significantly and positively related with land improvement, as fertilizer application increases soil carbon [34], which could correspond to an increase in NDVI. Increases in the population density would lead to land degradation but not significantly. According to the previous studies, the impact of population density on land degradation is ambiguous, while our results suggested that there would be more serious land degradation in areas with higher population density [35]. The population growth will lead to increasing demand for housing and other facilities, which in turn can lead to increase of impervious surface as a result of urban development and infrastructure construction and deforestation. Rapid economic development, which can be represented by changes in the per capita GDP, shows that an increase in both GDP and NDVI had positive impacts on land quality, suggesting the role the ecosystem service could play in economic growth and prosperity. Economic development would stimulate the land improvement, but such impacts were not significant, while the rural economic development, represented by rural farmers' per capita income, showed negative relationship with land improvement, suggesting that the NDVI will decrease as the rural farmer's income increase. It may be due to the fact that rural economic growth develops at the cost of intensive land use practice, which contributes to environmental degradation further leading to the land degradation. In addition, the increases in the share of the primary industry (agriculture, animal husbandry, and fishery) in GDP significantly corresponded to the decrease in NDVI. As production value of the agriculture increases mostly resulted from the investment into agricultural production, such as fertilizer, pesticide, and farm machinery, not from natural land quality improvement. In addition, the geographic and topographic factors also affect the land degradation. As to the topographical factors, the results showed that the higher the elevation and the slower the slope, the less serious the land degradation. The impact of elevation was not significant, while it may suggest that, in the areas with higher elevation, there would be less impacts of anthropogenic factors on the land degradation. The slope had clear significant impacts on land degradation, suggesting that steep slopes would lead to land degradation more easily, as steep slope regions were more vulnerable to severe water-induced soil erosion.

As to the soil fertility represented by soil organic matter, the results showed that the higher the soil fertility was, the less the possibility of land degradation is. Besides, the distance to highway was significantly positively related with land improvement, which suggested that the larger distances to road network meant that there was less human disturbance, less infrastructure development, and land use changes, which would exert less impacts on the land degradation.

CONCLUSIONS AND DISCUSSIONS

Land degradation or improvement is an outcome of many highly interlinked direct and underlying causes, including natural, socioeconomic, and related agricultural practices. In China, many types of land degradation are occurring, such as grassland degradation, deforestation, and cultivated land degradation. In this paper, we conducted an empirical study in the North China Plain and analyzed the basic grain production changes, the changing trend of NDVI, the spatial pattern of relationship between climatic factors and NDVI. Further taking NDVI as a proxy for land degradation and improvement, we applied an econometric model to analyze the causes of land degradation.

According to the analyses, both the grain production and grain yield of the North China Plain showed an increasing trend during the years 2000–2008; however, estimation based on the remotely sensed NDVI data showed that the North China Plain still experienced land degradation. Degraded areas have been expending in the northern part of the study area, though many parts of the study area showed land improvement. Based on the correlation analysis, the increases in temperature and rainfall corresponded to the increase in NDVI in most parts of the North China Plain, though some parts of the study area showed negative correlation between NDVI and temperature and rainfall. Further, according to the binary panel logistic regression analysis, the results showed there were strong impacts of some key biophysical and socioeconomic variables on land degradation. Some of these relationships were consistent with conclusions in previous case studies, while others showed complex differences. The results for climate factors were not surprising, and increase in temperature and rainfall would contribute to land improvement. Different land use changes exert different impacts on land degradation. Unsurprisingly, increasing of forest land and grassland would benefit land quality, while urban expansion would lead to land degradation. Our estimation results about the positive effects of agricultural intensification represented by fertilizer utilization on land degradation were also as expected, and the soil organic matter was also a crucial positive factor for land quality conservation. Rural economic and agricultural development may lead to land degradation. The rural economic and agricultural production growth exerted negative impacts on land quality, which means the development of rural economy and agricultural production led to land degradation. The increases in rural farmers' per capita income and primary production value ratio did not result in the improvement of land quality, which may be due to the overexploitation of land with insufficient investment into the land conservation. In this regard, an effective response to land degradation needs increasing incentives of farmers to conserve their cultivated land and improve their access to the knowledge and inputs that are required for proper conservation. Promotion of such land improvements should be a development policy priority. During the process to promote rural economic development, the governments should also focus on the monitoring and assessment of land quality and make measures to improve land quality, and such improvement measures should be designed together with farmers to meet their prior needs and use appropriate techniques according to the local economic and social conditions. In addition, increasing accessibility to infrastructures may also have negative effects on land quality. Infrastructure development is the basis for regional prosperity, and a booming economy will result in more construction of infrastructure. The expansion of basic infrastructure of transportation such as roads, railways, and airports can further take up land resources and further result in land overexploitation and degradation. The covering of the soil surface with impervious materials as a result of urban development and infrastructure construction is known as soil sealing. Sealing of the soil and land consumption are closely interrelated; when natural, seminatural, and cultivated land is covered by impervious surfaces and structures, this will degrade soil functions or cause their loss. To reduce the impacts of infrastructure construction on land quality, the local government should take the assessment of land degradation into consideration during the construction of infrastructures.

Based on the above analysis results, to achieve sustainable land management, climate changes should be monitored so as to make adaptation measures to mitigate the impacts of climate changes on land quality; along with the socioeconomic development, investments and better land management for improving land quality should be definitely encouraged through appropriate policy measures; human activities that change the land surface, such as infrastructure constructions, should be regulated on the basis of the assessment of impacts on land quality, and corresponding land conservation measures should be taken during the construction process.

ACKNOWLEDGMENTS

This research was financially supported by China National Natural Science Funds for Distinguished Young Scholar (Grant no. 71225005). The data support from the National Natural Science Foundation of China (Grant no. 41171434) and the National Key Program for Developing Basic Science in China (Grant no. 2012CB955700) was also greatly appreciated.

REFERENCES

1. S. Bajocco, A. de Angelis, L. Perini, A. Ferrara, and L. Salvati, "The impact of land use/land cover changes on land degradation dynamics: a Mediterranean case study," Environmental Management, vol. 49, no. 5, pp. 980–989, 2012. · ·
2. E. B. Barbier, "The economic determinants of land degradation in developing countries," Philosophical Transactions of the Royal Society B: Biological Sciences, vol. 352, no. 1356, pp. 891–899, 1997. · ·
3. M.V. Sivakumar and N. Ndiang›ui, Climate and Land Degradation, Springer, 2007.
4. E. Symeonakis, A. Calvo-Cases, and E. Arnau-Rosalen, "Land use change and land degradation in southeastern Mediterranean Spain," Environmental Management, vol. 40, no. 1, pp. 80–94, 2007. · ·

5. Millennium Ecosystem Assessment, Ecosystems and Human Well-Being: Desertification Synthesis, World Resources Institute, 2005.
6. WMO, Cliamte and Land Degradation, World Meteorological Organization, 2005.
7. Z. Bai and D. Dent, "Recent land degradation and improvement in China," Ambio, vol. 38, no. 3, pp. 150–156, 2009. · ·
8. X. Deng, J. Huang, S. Rozelle, and E. Uchida, "Growth, population and industrialization, and urban land expansion of China," Journal of Urban Economics, vol. 63, no. 1, pp. 96–115, 2008. · ·
9. F. Long, Introduction to the resources status in China-Land resources Jingchengnew.cn, 2013 (Chinese),http://www.jingchengw.cn/new/20130411/4927.htm.
10. H. Li, Z. Liu, L. Zheng, and Y. Lei, "Resilience analysis for agricultural systems of North China plain based on a dynamic system model," Scientia Agricola, vol. 68, no. 1, pp. 8–17, 2011. · ·
11. X. Deng, J. Huang, S. Rozelle, and E. Uchida, "Cultivated land conversion and potential agricultural productivity in China," Land Use Policy, vol. 23, no. 4, pp. 372–384, 2006. · ·
12. H. Long, G. Tang, X. Li, and G. K. Heilig, "Socio-economic driving forces of land-use change in Kunshan, the Yangtze River Delta economic area of China," Journal of Environmental Management, vol. 83, no. 3, pp. 351–364, 2007. · ·
13. G. N. Bastin, G. Pickup, and G. Pearce, "Utility of AVHRR data for land degradation assessment: a case study," International Journal of Remote Sensing, vol. 16, no. 4, pp. 651–672, 1995. · ·
14. K. J. Wessels, S. D. Prince, P. E. Frost, and D. van Zyl, "Assessing the effects of human-induced land degradation in the former homelands of northern South Africa with a 1 km AVHRR NDVI time-series,"Remote Sensing of Environment, vol. 91, no. 1, pp. 47–67, 2004. · ·
15. D. Singh, M. S. P. Meirelles, G. A. Costa, I. Herlin, J. P. Berroir, and E. F. Silva, "Environmental degradation analysis using NOAA/AVHRR data," Advances in Space Research, vol. 37, no. 4, pp. 720–727, 2006. · ·

16. Q. Jiang, X. Deng, J. Zhan, and S. He, "Estimation of land production and its response to cultivated land conversion in North China Plain," Chinese Geographical Science, vol. 21, no. 6, pp. 685–694, 2011. · ·
17. W. Shi, F. Tao, and J. Liu, "Changes in quantity and quality of cropland and the implications for grain production in the Huang-Huai-Hai Plain of China," Food Security, vol. 5, no. 1, pp. 69–82, 2013. · ·
18. C. J. Tucker, "Red and photographic infrared linear combinations for monitoring vegetation," Remote Sensing of Environment, vol. 8, no. 2, pp. 127–150, 1979. · ·
19. C. J. Tucker, H. E. Dregne, and W. W. Newcomb, "Expansion and contraction of the Sahara desert from 1980 to 1990," Science, vol. 253, no. 5017, pp. 299–301, 1991. · ·
20. M. Elhag and S. Walker, "Impact of climate variability on vegetative cover in the Butana area of Sudan,"African Journal of Ecology, vol. 47, supplement 1, pp. 11–16, 2009. · ·
21. E. Nkonya, N. Gerber, P. Baumgartner et al., ZEF Discussion Papers on Development Policy, 2011.
22. Z. G. Bai, D. L. Dent, L. Olsson, and M. E. Schaepman, "Proxy global assessment of land degradation,"Soil Use and Management, vol. 24, no. 3, pp. 223–234, 2008. · ·
23. F. Yin, X. Deng, Q. Jin, Y. Yuan, and C. Zhao, "The impacts of climate change and human activities on grassland productivity in Qinghai Province, China," Frontiers of Earth Science, vol. 8, no. 1, pp. 93–103, 2014. · ·
24. G. Zhang, X. Xu, C. Zhou, H. Zhang, and H. Ouyang, "Responses of grassland vegetation to climatic variations on different temporal scales in Hulun Buir Grassland in the past 30 years," Journal of Geographical Sciences, vol. 21, no. 4, pp. 634–650, 2011. · ·
25. J. R. G. Townshend, C. Justice, D. Skole, et al., "The 1km resolution global data set: needs of the International Geosphere Biosphere Programme," International Journal of Remote Sensing, vol. 15, no. 17, pp. 3417–3441, 1994. · ·
26. J. Chen, P. Jönsson, M. Tamura, Z. Gu, B. Matsushita, and L. Eklundh, "A simple method for reconstructing a high-quality

NDVI time-series data set based on the Savitzky-Golay filter," Remote Sensing of Environment, vol. 91, no. 3-4, pp. 332–344, 2004. · ·

27. J. Liu, M. Liu, D. Zhuang, Z. Zhang, and X. Deng, "Study on spatial pattern of land-use change in China during 1995–2000," Science in China Series D: Earth Sciences, vol. 46, no. 4, pp. 373–384, 2003.
28. J. Liu, Z. Zhang, X. Xu et al., "Spatial patterns and driving forces of land use change in China during the early 21st century," Journal of Geographical Sciences, vol. 20, no. 4, pp. 483–494, 2010. · ·
29. M. F. Hutchinson and P. E. Gessler, "Splines—more than just a smooth interpolator," Geoderma, vol. 62, no. 1–3, pp. 45–67, 1994. · ·
30. D. Xiangzheng, Simulation of Land System Dynamics, China Land Press, Beijing, China, 2008, (Chinese).
31. P. A. Hancock and M. F. Hutchinson, "Spatial interpolation of large climate data sets using bivariate thin plate smoothing splines," Environmental Modelling & Software, vol. 21, no. 12, pp. 1684–1694, 2006. · ·
32. X. Zhang, Q. Ge, and J. Zheng, "Impacts and lags of global warming on vegetation in Beijing for the last 50 years based on remotely sensed data and phonological information," Chinese Journal of Ecology, vol. 24, no. 2, pp. 123–130, 2005.
33. E. Nkonya, N. Gerber, J. von Braun, and A. de Pinto, "Economics of land degradation," IFPRI Issue Brief68, 2011.
34. P. L. G. Vlek, G. Rodríguez-Kuhl, and R. Sommer, "Energy use and CO_2 production in tropical agriculture and means and strategies for reduction or mitigation," Environment, Development and Sustainability, vol. 6, no. 1-2, pp. 213–233, 2004. · ·
35. S. J. Scherr, "A downward spiral? Research evidence on the relationship between poverty and natural resource degradation," Food Policy, vol. 25, no. 4, pp. 479–498, 2000. · ·

Chapter 7

Impact of Land Use Change on the Local Climate over the Tibetan Plateau

Jiming Jin[1], Shihua Lu[2], Suosuo Li[2,] and Norman L. Miller[3]

[1]Departments of Watershed Sciences & Plants, Soils, and Climate, Utah State University, Logan, UT, USA

[2]Cold and Arid regions Environmental and Engineering Research Institute, Chinese Academy of Sciences, Lanzhou, China

[3]Earth Sciences Division, Lawrence Berkeley National Laboratory, and Department of Geography, University of California at Berkeley, Berkeley, CA, USA

ABSTRACT

Observational data show that the remotely sensed leaf area index (LAI) has a significant downward trend over the east Tibetan Plateau (TP), while a warming trend is found in the same area. Further analysis indicates that this warming trend mainly results from the nighttime warming. The Single-Column Atmosphere Model (SCAM) version 3.1 developed by the National Center for Atmospheric Research is used to investigate the role of land use change in the TP local climate system and isolate the contribution of land use change to the warming. Two sets of SCAM simulations were performed at the Xinghai station that is located near the center of the TP Sanjiang (three rivers) Nature Reserve where the downward LAI trend is largest. These simulations were forced with the high and low LAIs. The modeling results indicate that, when the LAI changes from high to low, the daytime temperature has a slight decrease, while the nighttime temperature increases significantly, which is consistent with the observations. The modeling results further show that the lower surface roughness length plays a significant role in affecting the nighttime temperature increase.

INTRODUCTION

It has become apparent that climate change over high-elevation regions is occurring at a faster rate than over low-elevation regions [1]. This high-elevation rate of change often generates stronger disturbances within the climate system than low-elevation changes [2]. The Tibetan Plateau (TP) is an immense upland area (3500×1500 km), with an average elevation greater than 5,000 m. Over the TP, 46% of the forest cover and 50% of the grassland have been converted to farmland, urban areas, or desert during the last century (X. Liu, personal communication). The TP is the source of snowmelt runoff, supplying water resources to users in China and surrounding areas. The mechanisms influencing the observed rapid climate change and their impacts at this high-elevation region are not well understood. In order to reduce these types of uncertainties and provide society with more reliable projections of future outcomes, detection and analysis of these processes need to be determined.

Many researchers have shown that land use and land cover change has remarkab interacts with climate systems [3, 4]. A statistical analysis of the southern half of the Central Valley was performed by Christy et al. [5] using high-quality temperature observations for the San Joaquin Valley in California, where land use has changed dramatically since the presettlement (circa 1860s), due to extensive agricultural expansion. Their study indicated that the daily maximum near-surface air temperature exhibited a cooling trend over the last century for this region. This finding is in opposition to the observed warming trends that prevail over most other regions (see [6, 7]; many others), suggesting that agricultural activity and associated irrigation processes most likely play a key role in producing this cooling trend. Modeling studies have shown that land use change affects regional climate by altering the components in the surface energy budget [8]. Based on the results from a general circulation model (GCM), Snyder et al. [9] indicate that removing all temperate forest and replacing it with bare soil produce cooling in the winter and spring over tropical and boreal areas due to an increase in the surface albedo, but a warming during the summer due to reduced evapotranspiration. If forced with a moderate amount of carbon dioxide, GCMs could produce similar results for the entire 21st century [10]. These GCM predictions also indicate that the response of land use change could be overridden by a strong atmospheric circulation system such as the Asian Monsoon circulation.

However, due to the lack of detailed long-term and densely distributed spatial observations, researchers have not been able to determine how and to what extent these land use changes affect the early and accelerated warming observed on the TP [11]. A high-quality climate model is an efficient tool to advance our understanding of this problem, and observed climate change could be quantitatively interpreted through the model details. The modeling results would further yield new insight into future field experimental design. Therefore, in this study, we use the Single-Column Atmosphere Model (SCAM) version 3.1 coupled with the Community Land Model version 3 (http://www.ccsm.ucar.edu/models/atm-cam/docs/scam/) to investigate the role of land use change in the TP local climate system and isolate the contribution of land use change to the warming of the TP from that of greenhouse gas emissions. The SCAM was developed by the National Center for Atmospheric Research (NCAR). In addition, the mechanisms associated with land use change signals are adequately quantified with the model output and our existing quality-controlled observations.

MODEL, DATA, AND METHODOLOGY

The SCAM used in this study is a one-dimensional time-dependent single-column atmospheric model. The SCAM is embedded in the NCAR Community Atmospheric Model version 3. Since the SCAM includes only one model grid cell, it can be used to effectively study land atmosphere interactions. This grid cell can include up to 10 land use types, giving a benefit to characterize land surface heterogeneity. In this study, the land use types are prescribed in the SCAM to represent land surface features in the TP. The SCAM is also configured with 26 vertical layers ranging from approximately 3 mb to the surface, and it includes all the physical schemes existing in its global counterpart, CAM3. The reanalyzed data by National Centers for Environmental Prediction (NCEP-R1) provide the initial and lateral boundary conditions for SCAM. The latter were updated every 6 hours. The initial conditions include soil moisture and temperature and atmospheric pressure, moisture, temperature, and winds. The 16 km resolution Advanced Very High Resolution Radiometer (AVHRR) leaf area indices (LAIs) were used to examine vegetation changes over the TP and also were inputted into SCAM to more realistically reflect vegetation variations. Station observations were used to analyze climate change over the TP and compare with modeling results.

RESULTS

LAI, Temperature, and Precipitation Analysis

Land use and land cover on the TP have been changed significantly. Such change can be indentified with the AVHRR LAI trends over the period of 1983–2000 for July, August, and September, a season that the LAIs usually reach their maximum values (Figure 1). It is clearly seen that the LAIs have negative trends (blue colors) in the northeast TP, and the maximum trend can reach around −1.0/decade. These trends are calculated through a linear regression method. The region with the negative trends contains the headwaters of the Yellow River (Huang He), the Yangtze River (Chang Jiang), and the Mekong River

(Lancang Jiang) that provide a significant amount of water resources for China. This area is also called Sanjiang (three rivers) Headwaters Nature Reserve. In some locations of the southeast TP, LAIs also show positive trends (red colors). Figure 2 shows the time series of the annual maximum LAI at a station called Xinghai that is located near the center of the Sanjiang area. This figure shows that the annual maximum LAI has a downward trend with a value of −0.67/decade over the period of 1983–2000. According to the United States Geological Survey data, Xinghai has a flat terrain and mostly is covered by grassland during the summer. All the following simulations were performed at this station to understand how vegetation affects local climate.

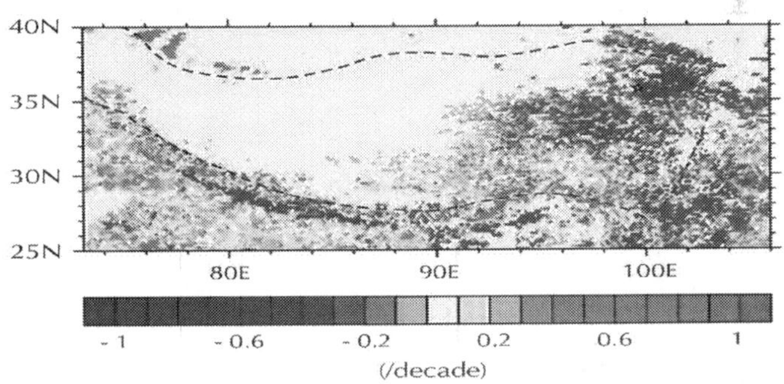

Figure 1: The LAI trend in the TP over the period of 1983–2000 (unit: /decade). The dashed line represents the scope of the TP. The black star is the location of the Xinghai station where the SCAM simulations were performed.

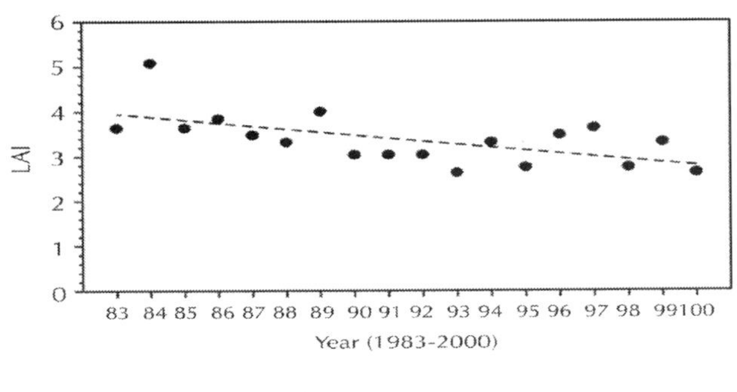

Figure 2: The trend of the annual maximum LAI at the Xinghai station for the period of 1983–2000. The dashed line represents the trend. The black dot is the maximum LAI for each year.

Figure 3(a) shows the time series of the annual precipitation anomalies at the Xinghai station for the period of 1960–2000. It is shown that the trend is 0.5 mm/decade that is quite insignificant. However, the annual mean temperature trend is 0.25°C/decade (Figure 3(b)). The temperature increase is about 1°C at this station over the study period. When temperatures at 2 AM and 2 PM are examined, it can be seen that the nighttime temperature trend is 0.35°C/decade (Figure 3(c)) while the daytime temperature trend is only 0.04°C/decade (Figure 3(d)). Thus, the annual mean temperature increase results from the nighttime warming. Identifying physical factors and processes that contribute to the nighttime warming is the major focus in the following sections, and the SCAM is used for such purpose.

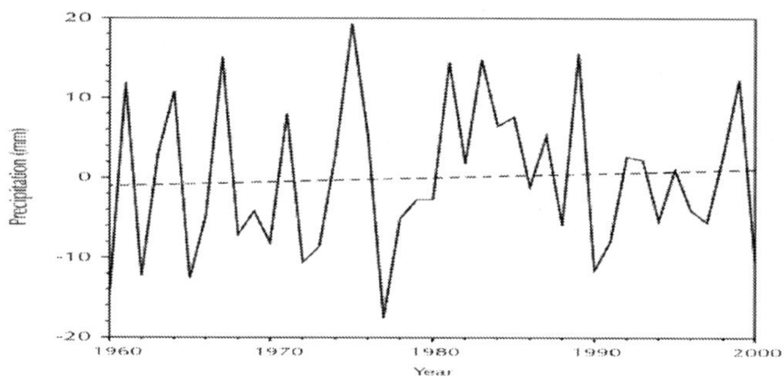

(a)

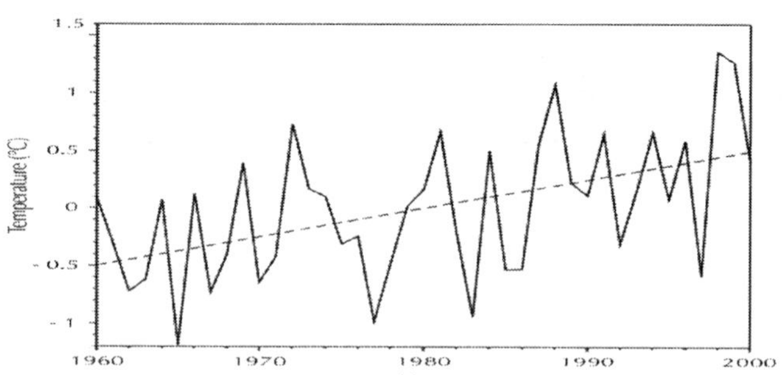

(b)

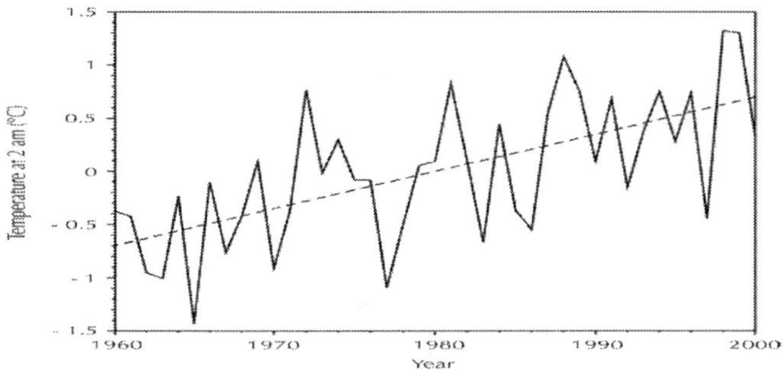

(c)

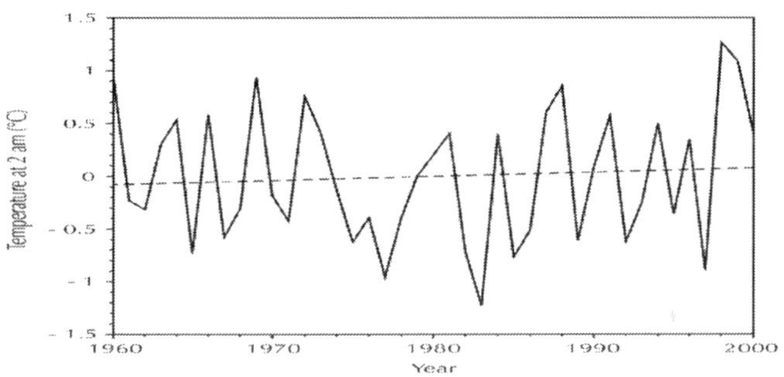

(d)

Figure 3: The time series of annual precipitation and temperature anomalies at the Xinghai station. (a) Precipitation (mm); (b) temperature (°C); (c) temperature at 2 AM (°C); temperature at 2 PM. The dashed line in each figure represents the trend.

SCAM Evaluation

Before physical processes that affect temperature change are studied, the SCAM needs to be fully evaluated with observations to ensure that it can reasonably describe climate processes in our study area. SCAM simulations at the Xinghai station were performed over the period of 1951–2000. The first 32-year (1951–1982) simulations were discarded for spinning up the model, and only the simulations over the period of 1983–2000 were analyzed. Figure 4(a) shows that although the SCAM produces a weaker seasonal cycle of the temperature it still generates reasonable temperature simulations at the Xinghai station when compared to the observations. The correlation coefficient between the observation and the simulation is 0.98, the RMSE is C, and the bias is 0.29°C. The model can also reproduce the observed precipitation, where the correlation coefficient, RMSE, and bias are 0.74, 28 mm/month, and 13 mm/month, respectively (Figure 4(b)). These results show that the SCAM can simulate well for the seasonal cycles of precipitation and temperature at the Xinghai station, indicating that this model can be used to further explore physical processes that affect temperature and precipitation changes in this area.

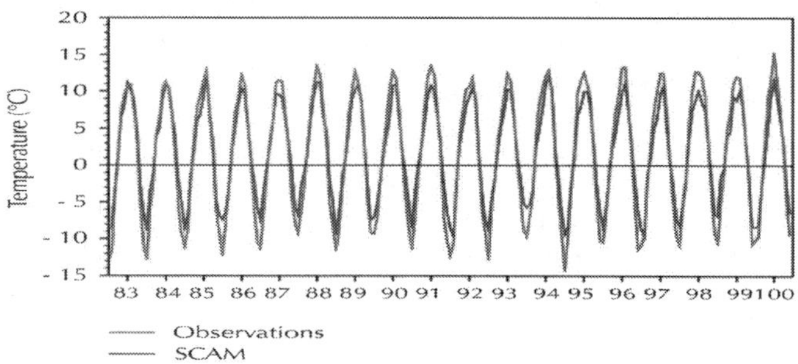

(a)

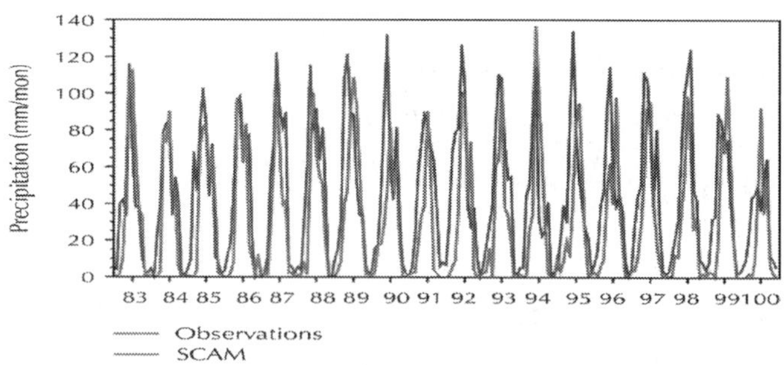

(b)

Figure 4: Comparison of SCAM simulations (blue line) and observations (red line) at the Xinghai station: (a) Precipitation; (b) temperature.

Sensitivity Studies for Understanding the Temperature Changes

To examine how vegetation affects the warming in this area, two additional sets of simulations were performed with high and low LAIs for 1983–2000, a period where the remotely sensed LAIs are available. Figure 5 shows seasonal variations of high and low LAIs. The high LAIs are the average over 1983–1985, while the low LAIs are the average over 1998–2000. The largest difference between the high and low LAIs is more than one. The first set of simulations was forced with the high LAIs, where the LAI only varied seasonally but did not have interannual variations. The modeling settings for the second set of simulations were exactly the same as those for the first set of simulations but with the low LAIs. Figure 6(a) shows the 2 m height temperature difference between the high LAI and the low LAI cases averaged over the 18-year period (1983–2000). It is seen that the major changes in temperature (low LAI minus high LAI) occur in the warm season where the LAI difference between the two cases is the largest. During the cold season (from October through next May), there is almost no change for the temperature. Precipitation has very minor changes with a maximum difference of less than 1 mm/month (figure not shown). Figure 6(a) indicates that the lower LAI in this region generates a slight cooling during the summer at noontime with a maximum temperature change of -0.2°C in July (solid line). All the components in the energy balance equation are examined. The results show that the sensible heat flux has most significant changes that affect 2 m height temperature (Figure 6(b)). It is seen that sensible heat flux decreases by 10 W/m^2 at noontime during the summer (solid line). Further analysis indicates that such a decrease results from the lower roughness in the low LAI case. In SCAM, the roughness length for vegetation in this area is 0.06 m, and for bare soil, it is 0.024 m. The low LAI reduces the areal weight of vegetation in the model grid cell, but increases the weight of bare soil when compared to the settings in the high LAI case. Thus, on average, the low LAIs generate a flatter surface than the high LAIs, which decreases the upward sensible heat flux during daytime and slightly lower the near-surface temperature.

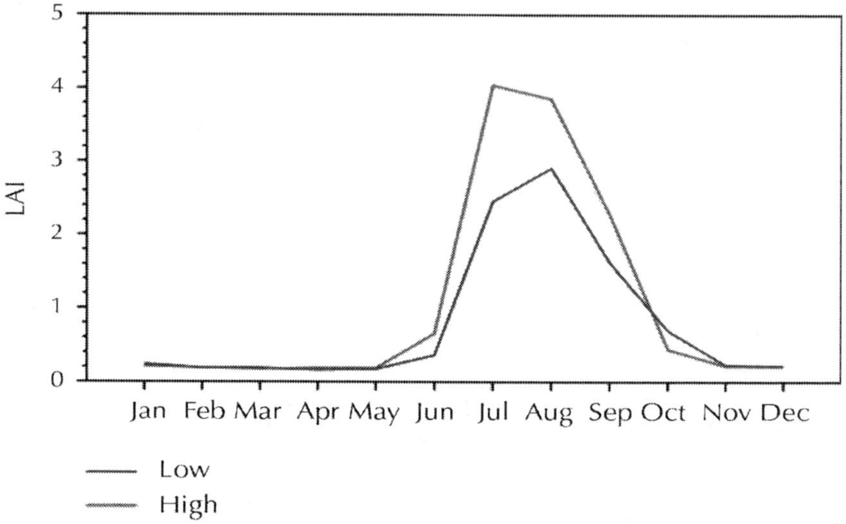

Figure 5: The seasonal variations of the low (blue) and high (red) LAIs. The high LAI is the average over 1983–1985, while the low LAI is the average over 1998–2000.

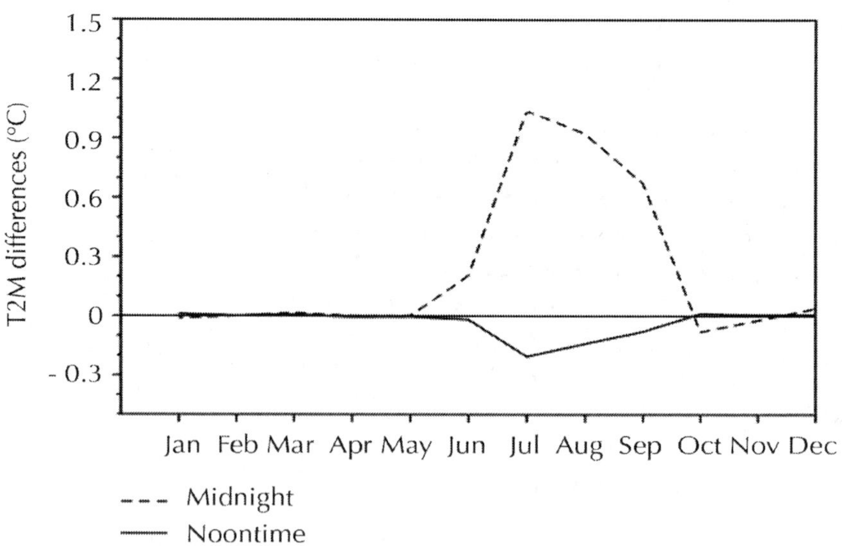

(a)

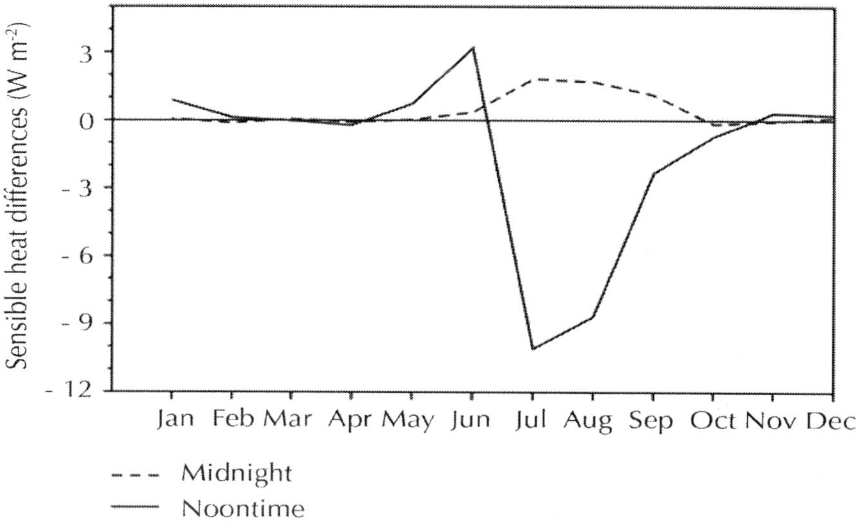

(b)

Figure 6: The (a) temperature (unit: °C) and (b) sensible heat flux (unit: Wm^{-2}) differences between the low and high LAI cases (low-high). The solid line is for noontime, and the dashed line is for midnight.

Figure 6(a) shows the nighttime temperature difference between the low and high LAI cases (low-high), indicating that the near-surface temperature increases by more than 1°C in July due to changing the LAI from high to low (dashed line). During the evening, the sensible heat flux is downward (negative). The reduced roughness length in the low LAI case suppresses the energy exchanges between the surface and near-surface air. Figure 6(b) shows that the increase in sensible heat flux at midnight is up to 2 Wm^2 (dashed line) that leads to more energy remaining in the near-surface air (less downward sensible heat) and increases the temperature there. Detailed calculation of sensible heat flux and how it is related to the roughness length are described by Oleson et al. in [12].

The above modeling results strongly suggest that the warming results from nighttime temperature changes that are consistent with the observations. The decreased vegetation LAI in this area significantly contributes to the nighttime temperature increase. The LAI decrease may not be a controlling factor that produces the observed warming trend, but most likely it reinforces the existing warming trend that is believed to be triggered by the carbon increase in the atmosphere [4].

CONCLUSIONS AND DISCUSSIONS

In northeast Tibetan Plateau, the remotely sensed LAIs show a significant downward trend, while a warming trend is observed for temperature. Our modeling results indicate that such a vegetation change is an important factor that contributes to the warming in this region. Further analysis indicates that this contribution largely comes from the evening temperature increase. The evening temperature increase results from the more stable near-surface boundary layer that restricts the downward sensible heat flux and keeps more energy in the near-surface air. The deterioration of vegetation leads to a flatter area, reduces the surface roughness, and results in more stable near-surface boundary layer.

This study is not intended to indicate that vegetation deterioration is a determining factor that leads to a warming in this area that is most likely related to global climate change triggered by the atmospheric carbon increase. Instead, the analysis shown in this study describes a fact that the reduced biomass in this area could strengthen the warming trend in this region. This study is based on the modeling results that need to be further verified with more observed evidence. The large- and mesoscale atmospheric information in the SCAM was introduced through the lateral boundary data. Thus, the atmosphere in the model column does not produce any feedback to that outside the column. This drawback could be overcome by applying a regional or global climate model to this region, which could be a future research topic. However, these modeling results give strong insight into future modeling and observational studies for climate change in this high-elevation area.

ACKNOWLEDGMENTS

Support for this work was provided by the Utah Agricultural Experiment Station. J. Jin was also supported by K. C. Wong Fellowships, Chinese Academy of Sciences.

REFERENCES

1. H. F. Diaz and R. S. Bradley, "Temperature variations during the last century at high elevation sites,"Climatic Change, vol. 36, no. 3-4, pp. 253–279, 1997.
2. J. Jin and N. L. Miller, "Toward understanding the role of ground water in hydroclimate over the Merced watershed using a single column climate model," in The American Geophysical Union Fall Meeting, San Francisco, Calif, USA, December 2006.
3. Q. Li and Y. Xue, "Simulated impacts of land cover change on summer climate in the Tibetan Plateau,"Environmental Research Letters, vol. 5, no. 1, Article ID 015102, 2010.
4. X. Cui, H.-F. Graf, B. Langmann, W. Chen, and R. Huang, "Climate impacts of anthropogenic land use changes on the Tibetan Plateau," Global and Planetary Change, vol. 54, no. 1-2, pp. 33–56, 2006.
5. J. R. Christy, W. B. Norris, K. Redmond, and K. P. Gallo, "Methodology and results of calculating central California surface temperature trends: evidence of human-induced climate change?" Journal of Climate, vol. 19, no. 4, pp. 548–563, 2006.
6. J. Hansen, M. Sato, R. Ruedy, K. Lo, D. W. Lea, and M. Medina-Elizade, "Global temperature change,"Proceedings of the National Academy of Sciences of the United States of America, vol. 103, no. 39, pp. 14288–14293, 2006.
7. J. T. Houghton, Y. Ding, D. J. Griggs, et al., Climate Change 2001: The Scientific Basis: Contribution of Working Group I to the Third Assessment Report of the Intergovernmental Panel on Climate Change, Cambridge University Press, 2001.
8. L. M. Kueppers, M. A. Snyder, L. C. Sloan et al., "Seasonal temperature responses to land use change in the western United States," Global and Planetary Change, vol. 60, no. 3-4, pp. 250–264, 2008.
9. P. K. Snyder, C. Delire, and J. A. Foley, "Evaluating the influence of different vegetation biomes on the global climate," Climate Dynamics, vol. 23, no. 3-4, pp. 279–302, 2004.
10. J. J. Feddema, K. W. Oleson, G. B. Bonan et al., "Atmospheric science: the importance of land cover change in simulating future climates," Science, vol. 310, no. 5754, pp. 1674–1678, 2005.

11. O. W. Frauenfeld, T. Zhang, and M. C. Serreze, "Climate change and variability using European Centre for Medium-Range Weather Forecasts reanalysis (ERA-40) temperatures on the Tibetan Plateau,"Journal of Geophysical Research D, vol. 110, no. 2, pp. 1–9, 2005.
12. K. W. Oleson, Y. Dai, G. Bonan, et al., "Technical description of the community land model (CLM),"Technical Note NCAR/TN-461+STR, NCAR, 2004.

Chapter 8

An Agent-Based Reasoning of Impacts of Regional Climate Changes on Land Use Changes in the Three-River Headwaters Region of China

Tao Zhang[1], Jinyan Zhan[1], Juan Huang[1], Rui Yu[2], and Chenchen Shi[1]

[1]State Key Laboratory of Water Environment Simulation, School of Environment, Beijing Normal University, Beijing 100875, China

[2]Faculty of Resources and Environmental Science, Hubei University, Hubei, Wuhan 430062, China

ABSTRACT

The land surface in Three-River Headwaters Region (TRHR), a typical ecological fragile zone of China, is quite sensitive to the climate changes which will destabilize certain ecosystem services valuable to the entire nation and neighboring countries. This study aimed to analyze the impacts of climate changes and agents' adaptive behaviors on the regional land use changes with the agent based model (ABM). First, the main agents were extracted according to the production resources endowments and socioeconomic background. Then the agents' land use behaviors were analyzed and parameterized. Thereafter, the ABM model was built to simulate the impacts of the climate changes on the regional land use changes and agents' economic benefits. The results showed that the land use changes were mainly characterized by the increase of grassland and decrease of unused land area. Besides, the agents would get more wealth under the scenario without climate changes in the long term, even though the total income is lower than that under the scenario with climate changes. In addition, the sensitivity analysis indicated that the model is sensitive to the climatic conditions, market price of agricultural and animal husbandry products, government subsidies, and cost control.

INTRODUCTION

Both the global and regional climate changes have greatly undermined the terrestrial landscapes, ecological processes, and ecosystem services [1–4], which have subsequently threatened not only the human societies but also the natural environment itself by the recondite feedback effects [5–8]. There have been dramatic changes in the climatic conditions of China, which will continue to last for a long time in the future. The atmospheric temperature of China has increased by 0.5–0.8°C in the 20th century, with the upper limit exceeding the global average level. Meanwhile, the precipitation has continuously decreased in northern part of China while increased greatly in the southern and southeast part of China. Additionally, the extreme climate and subsequent natural disasters have also showed a longitudinal distribution pattern. A number of researchers have reported the impacts of climate changes on various factors of the ecological and social systems [9, 10], such as terrestrial surface [10–12], the market price [13, 14], and externality of public goods and even the political negotiations among various stakeholders [15, 16].

Climate changes associated with human adaptive activities, especially the social agents' decisions on land use, have exerted synergistic impacts on the land surface at different scales [11, 17, 18].

There are generally two approaches through which the climate changes influence the dynamic land surface properties, that is, the natural processes and human adaptive behaviors. The natural processes, for example, the change of thermodynamic properties and hydrological changes, will alter the terrestrial land surface at the long-term scale, and the inherent properties of different land cover types limit the land use activities of human beings, while the human adaptive behaviors may mitigate the adverse effects of climate changes through changing the land use practices. For instance, the herdsmen prefer pasturing sheep on the grassland to other land cover types, and therefore they may transform the cultivated land and forest land into grassland so as to adapt to the climate changes and obtain more ecosystem services. However, there has been very limited knowledge on the impacts of climate changes on the land use dynamics through disturbing microagents' behaviors so far. Besides, the climate change has also influenced the behaviors of government agents, who have taken measures to mitigate the impacts of the climate change and may subsequently influence the behaviors of other agents. For example, the long-term strategic development policy in the western region of China has greatly promoted the ecological restoration and protection of the regional ecosystem; the subsequent national wide policies of Grain for Green (GFG) and Overgrazing Forbidden have made remarkable progresses in the land planning and ecological construction through compensating farmers and herdsmen for the change of their unreasonable land use practices; what is more important is that the establishment of the Three-River Headwaters Region Nature Reserve has imposed restrictions on the local human land use practices and consequently protected the local ecological environment.

It is necessary to analyze the driving mechanism of social agents' behaviors and clarify their behaviors or decision algorithms with a proper model in order to more accurately study the temporal and spatial transitions of land use under the impacts of the climate change and social agents' behaviors. The agent-based model (ABM) is a mainstream model to specify the behaviors of agents. Although there have been many ABM studies on the regional and local simulation, especially in the urban areas and transition zones between cultivated land and grassland, it is still necessary to carry out more in-depth ABM studies in areas without significant land use and land cover change (LUCC) and easy-distinguished agents. Besides, previous researches on the identification and parameterization of the agents' behaviors have mainly relied on the mathematical or econometric theories, which are different from the thinking ways of human beings from the ethological perspective in the real life. What is more, agents will actually choose the land use practices for a targeted land parcel according to their own or neighbors' successful land use experiences, which has served as the theoretical basis of the case based reasoning (CBR) in this study.

The Three-River Headwaters Region (TRHR), covering a total area of 37.2×10^5 km^2, is located in the south of Qinghai Province, China, and it accounts for 51.9% of total provincial area. The annual average temperature and precipitation in this area are 3.96°C and 500 mm. respectively, with an annual increment of 0.04°C and 0.17 mm. The grassland is the dominant land use type in this area, accounting for 71.5% of the study area and having a total decrement of 0.03% in the past two decades, while the unused land accounted for 16.5% of the study area and increased by 0.10% in the same period (Figure 1). The high and mediate coverage grasslands have experienced continuous shrinkage, while the low coverage grassland area has had continually expanded. The spatial pattern of land use changes in the past decades was characterized by the grassland degradation in the middle and east regions and the shrinkage of the water body in the north and west regions [19]. Besides, the economy in the study area has experienced unprecedented development during the past 20 years. The gross domestic product (GDP) in 2010 is 15 times of that in 1990. The animal husbandry is the dominant sector in the primary industry of this region, appropriately accounting for 79.5% of the total output of the primary industry.

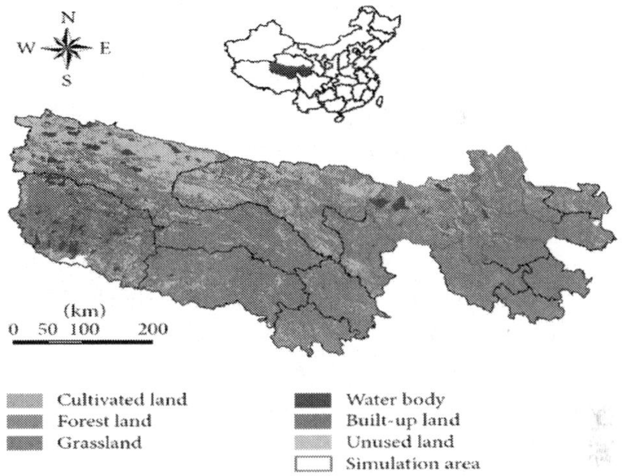

Figure 1: The location of study area. The simulation area, a continuous region involving all the land use types, was selected to implement the simulation with the ABM model. This figure shows the land use map in 2005.

This study has aimed to analyze the impacts of climate changes and variability and corresponding agents' adaptive land use decisions on the land surface in the TRHR with the agent-based model. This paper will firstly extract the main agents based on the analysis of production resources endowments and socioeconomic situation in the study area and then parameterizes the behaviors of various agents by utilizing county level statistical data and household survey data. Thereafter the ABM model was constructed and used to simulate the land use changes in the next 50 years. The economic benefits of agents are also discussed in this paper.

DATA AND METHODOLOGY

Data Collection

The data used in this study mainly includes the land use data, socioeconomic and demographic information, the natural environment and climatic conditions, all of which were derived from the following sources: (1) Environmental Science Data Center of Chinese Academy of Sciences, which offers the datasets of land use, digital elevation model (DEM), topography, the distance to water and road, and climate and soil properties; (2) statistical yearbooks, such as Qinghai Province Statistical Yearbook, China Agricultural Yearbook, and China Animal Husbandry Yearbook, which mainly contain the population, GDP, food production, and animal husbandry production; (3) household survey data, which were obtained from the household survey implemented by the research team in August and October 2012 with the stratified random sampling method performance. Three counties were selected as the sample sites to conduct the survey, which aims to identify the agents and parameterize their behaviors. The questionnaires have been modified after the pretest. Besides, an intraday examination was carried out on the result of questionnaires so as to revise the inconsistent or unreasonable answers with the retelephone investigation. This could not only save a lot of human resources and finance but also ensure the representativeness and reliability of the samples.

Identification of Agents

It is more difficult to identify the agents in the study area than in other regions where the agents such as farmers and herdsmen have inherent differences. In the study area, most people are herdsmen who rely on the grassland, and it seems that there are no absolute differences among these individuals. However, the statistical analysis and preinvestigation indicate that the herdsmen are different in the economic background and production resources, which would influence their land use decisions. Therefore, based on the analysis of the socioeconomic background, natural environment conditions, and resources endowments, all the counties in the study area were classified into three categories, that is, the traditional development area, animal husbandry dominated area, and restricted development area. By doing this, the major agents can be extracted from each category, including the farmer agent, herdsman agent, and hominid agent.

The farmer agents are mainly located in the traditional development areas, eastern regions. They possess abundant labor force, production resources, and relatively large cultivated land area, which help them get a higher economic return from diverse income sources. The herdsman agents mainly dwell in the animal husbandry dominated areas, where the cultivated land shares a very small proportion, but the grassland occupies more than 70% of the total area of these regions. So the herdsmen extremely rely on the grassland to develop the animal husbandry industry to get the revenue for supporting their families. Subsequently, the proportion of animal husbandry output to the primary industry output is higher than other sectors due to the intensive resources utilization and collective land use practices. The hominid agents were identified according to the habitat properties. The hominid agents were located in the counties where the social and natural ecological systems are extreme fragile. These counties are characterized by the harsh climatic conditions, lower economic levels and human resource, and lack of production resource endowments. Therefore, the agents in these areas would heavily rely on the natural conditions rather than the land use activities to support their life.

Agents' Behaviors

The general behavior rules of how the agents decide the land use practices for a targeted land parcel were introduced first, and then the individual behavior rules that varied from each agent were designed and specified with the diverse characteristics. At last, the impacts of the government agent's behaviors on other agents' decision through punishment and compensation were described.

Generally, all the agents have the same procedures of determining the land use practices for a certain land parcel. It means that each agent will choose a land use type for a land parcel at the beginning of each simulation step; then the agent will calculate the maximum combined economic and social benefit at the end of each step to decide which land use type should be chosen for this land parcel. Thereafter the agent chooses a certain land use type according to the land use decision algorithm. Finally, the agent will decide whether or not to sell or buy land parcels based on the estimation of economic account [20].

The individual behavior rules are different among agents due to the variation of economic background, natural conditions, and production resource endowments, which are vital to the design of the land use decision algorithm. There are three categories of individual decision algorithms in terms of satisfaction, imitation, and case-based reasoning (CBR) strategies [21]. It was assumed that agents have different probabilities to choose the decision algorithm according to the results of the field survey. Specifically speaking, if the benefits of the land use decision have reached the agents' expectation, the agents will choose the satisfactory land use strategy, that is, choose the same land use type in last year. Or else the agents will get certain probabilities to choose the imitation strategy or CBR strategy if the expectation is not realized [21]. As for the imitation strategy, the agents will imitate the successful land use practices of their neighbors. But actually all the agents are more willing to determine the land use for the land parcel according to their own historical experiences. Therefore, they will get a relative higher probability to choose the CBR strategy.

The CBR strategy is an approach to acquire the solutions to land use decisions through analyzing the historical experience. It can simplify the knowledge, improve efficiency, and update the cases [20]. Each case records one agent's land use experiences in a specific land parcel, which includes the year, biophysical characteristics of the land parcel, climatic conditions, land use type, and economic and social benefits. Cases in the model can be sorted and updated through putting the latest case in the forefront of the retrieval store when running the model [21]. The agents will retrieve the case store to find the most appropriate case that matches the properties and conditions in the simulation year. By doing this, the agents can effectively and quickly collect detailed information such as the economic and social benefits and land use type to decide whether or not to choose this land use type. Actually, as for the utilization of CBR to describe the agents' behaviors, the memory capacity of the agents is limited by the memory time and size. It is assumed that the agents can always remember several latest cases but not forever. And what is more important is that the excessive cases will decelerate the retrieval speed and require advanced hardware configurations.

The government agent relies on the participation message interface to pass parameters in the model, so it is also necessary to conduct quantitative and positioning analyses. In the ABM model, it is assumed that the government agent pays more attention to the ecological and environmental problems, and the government agent's behaviors were parameterized on the basis of the ecological compensation policies and land planning. With regard to the ecological compensation policies, a specific analysis was conducted with the Grain for Green policy as an example. The Grain for Green policy insists that the government should offer some compensation for the conversion from cultivated land to forestry area or grassland, and there is no restriction on the total amount of compensation fee to practitioners, given that the amount of compensation fee is just related to the total land conversion area. Besides, the nature reserves and future land planning will be considered, all of which are the bases to support the parameterization of the government agent's behaviors.

Key Factors of Agents' Behaviors

Table 1 shows the key influencing factors of the agents' behaviors, such as socioeconomic background, bad events, and human awareness to society and natural conditions. Since these factors can trigger the agents' behaviors and destabilize the magnitude of activities, different parameters of these factors were used to specify the land use decisions of agents.

Table 1: The factors that impact the agents' behaviors

Categories	Variables	Impacts on behaviors
Climate	Temperature Precipitation	Different climatic conditions determine the agents' land use practices; for example, mild climate is suitable for cropping, pasturing, and so forth.
Economy	Initial account Nonagricultural income Price	The amount of the agents' account impacts their land use practice, such as investment, land bid, and cost control.
Bad events	Natural disasters	Weight of balancing the economy and nature.
Policy	GFG Nature reserve	The compensation can stimulate the agents to transform other land uses into grassland and forest land.Ban the agents' land use practices in the core area.
Topography	Distance to water Slope	The conversion area of grassland to cultivated land decreases with the increment of distance to water and slope.

Soil property	Soil nitrogen Soil phosphorous Soil potassium	Agents are willing to keep the fertile land parcel as grassland.
Land use	Land use	The initial land use determines the results of next step.The land use type and area determine the probabilities of land use change.

The total economic account of agents, which comprises of initial account, agricultural income, nonagricultural income, and government compensation, can influence the agents' behaviors through the bid amount of land transformation and market prices of agricultural and animal husbandry products. The agents who own more wealth (higher purchasing power) are more willing to buy land parcels in the land transformation step (higher bid amount). The statistical data indicated that there is a difference in the initial accounts of agents, and the nonagricultural income of the farmer agents is higher than other agents. Besides, the agricultural income depends on the production and market prices; therefore it is assumed that the agents are willing to change the cultivated land into grassland if the prices of animal husbandry products are higher than that of the agricultural products. In addition, there is also difference in the land use properties of different agents, which will further lead to the difference in their behaviors. Generally, the land use of last simulation step determines the result of the next simulation step. For example, if one agent chooses the satisfactory strategy according to the land use algorithm, the land use of last year will be selected for the targeted land parcels. Moreover, small land parcels are relatively easier to be converted, and therefore their land conversion probabilities would be higher than plots within larger areas, and the agents' wealth and the area of each land use type are not the same; different land use types with different areas require diverse investments and cost. Therefore, the configuration of relevant parameters are different among these agents and land use types. What is more, there are still some other factors that may influence the agents' decisions through destabilizing their balance analysis between two-dimensional benefits of economic returns and social acceptance, such as the initial human awareness and natural disasters. For example, agents who have large population and high economic revenue and lack the needed ecosystem services may pay more attention to the social benefits and ecosystem services rather than the economic return. The agents who suffered from natural disasters may also reanalyze the two-dimensional benefits.

Parameterization and Validation

The parameters of the ABM model mainly include the model parameters and agents' parameters. The model parameters include the topology, neighbor radius, and cell numbers and sizes, all of which are the same for all the kinds of agents. These model parameters were mainly set on the basis of the statistical data at the county level, while the agents' parameters were primarily set according to the results of county classification and household survey data. In addition, the time delay of land use was taken into account in this study. There is some time delay in the impacts of the social agents' adaptions and government policies on the land use change. The agents cannot immediately change the land use practices of targeted parcels even if the income loss appears; it will take them some time to make sure whether they will change the land use practices or not. Therefore the parameter of the model was adjusted on the basis of the time delay.

The land use structure and pattern in 1995 and 2005 were first simulated on the basis of land use data of 1988 and then compared with the actual land use data so as to validate and calibrate the simulation model. The validation methods include the point to point comparison, Kappa index, Moran I index, and ROC curve. The results of all the tests showed that the ABM model has a good simulation capacity, but the result of the ROC test indicates there are also some potential uncertainties. The ROC curve showed that the simulation results of cultivated land and built-up land were not as good as those of other land use types, which may be due to the rough simulation resolution. The ABM model was finally calibrated through changing the sensitive parameters on the basis of the trials and errors.

RESULTS AND DISCUSSION

This study aims to explore the impacts of climate changes on the land use changes and the agents' economic benefits in the Three-River Headwaters Region. The land use changes were simulated under two scenarios, one with the climate changes and the other without the climate changes.

Land Use Changes

Most land use types except the cultivated land will change in a similar trend under both of the two scenarios. The simulation result suggests that the grassland and the built-up land will show an increasing trend, while the forest land and unused land will tend to shrink in the simulation period. On the whole, the area of grassland and unused land will change most greatly. More specifically, there is obvious change in the grassland area, with an increment rate of approximately 5.1% under the scenario with climate changes and 6.0% under the scenario without climate changes. On the contrary, the areas of forest land and unused land will decrease during the same period. The area of forest land will decrease by 23.4% and 33.7% and under the scenario with and without climate changes, respectively, while that of the unused land will be 14.6% and 8.7%, respectively (Figure 2). Therefore, the climate changes will lead to the decrease of cultivated land and unused land and the increase of forest land and grassland by comparing the results of these two scenarios on the whole.

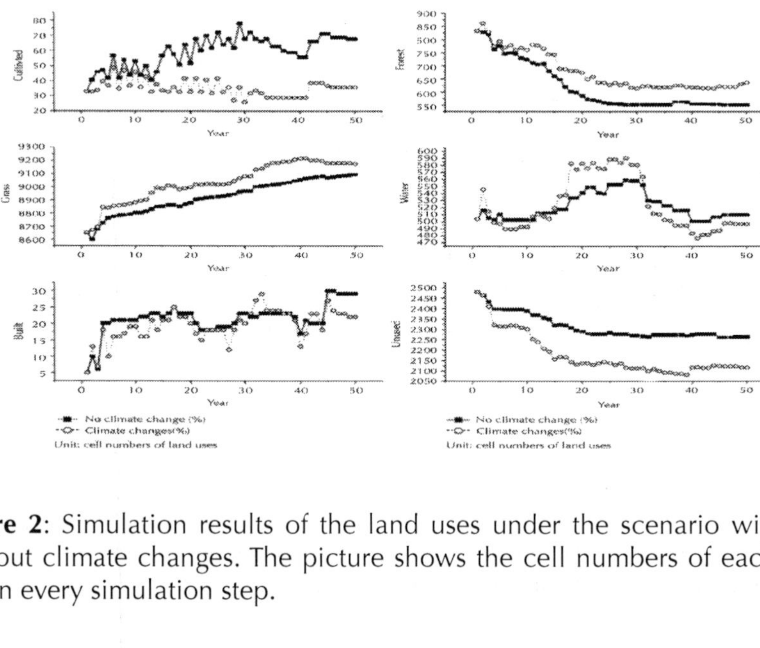

Figure 2: Simulation results of the land uses under the scenario with and without climate changes. The picture shows the cell numbers of each land use in every simulation step.

The area of the cultivated land, water body and built-up area will fluctuate strongly under the two scenarios. The cultivated land area will decrease to some degree under the scenario with climate changes and show an opposite trend under the other scenario. However, there are similar changing trends of the area of the water body and built-up land under these two scenarios but with more wide fluctuation range under the scenario with climate changes. This may be due to the rough simulation resolution. The land parcel was set as the basic unit for land use decision, which means that the agent will change the land use practices of all the land cells in the targeted land parcel. However, there were initially only 5 built-up land cells in the 2 km resolution data, and the smallest land parcel in the study area consists of 9 grid cells; as a result the resolution of the data leads to some errors in the simulation result. Besides, another reason may be the probability of climate change and the parameterization of the agents' behaviors, which account for the fluctuation of land use types in the large area. The probability of climate changes will influence the output of land use, which will further influence the agents' income, and therefore the agents will change their land use practices to adapt to the climate changes. In addition, there may be amplified impacts of climate changes on the land use changes on condition that the parameters of the agents' behaviors were not set reasonably.

Regional Economy

The change of the economic return was also simulated with the ABM model under different climate change scenarios. The economic return of agents mainly comes from three sources, that is, the agricultural income from the land parcel harvest, nonagricultural income, and government subsidies. The simulation results show that the climate changes will lead to a stable growth of the agents' wealth in the long term, and the agents' income will be higher under the scenario without climate changes, showing an exponentially increasing trend.

The herdsman's annual income is more than that of the other two agents since they own more grassland, from which the major income of the animal husbandry comes (Figure 3). Besides, the annual income of all agents showed a more optimistic changing trend under the scenario without climate changes. The change of the annual income of the farmer agents is similar to that of the hominid agent, showing an exponentially increasing trend. The annual income of the farmer agent will show a growing trend during the initial 25 years, followed by a declining trend under the scenario with climate changes. By contrast, the annual income of the hominid agent will be relatively more stable. In general, the total income of agents under the scenario without climate changes is lower than that under the scenario with climate changes (Figure 4). However, the total income of the farmer agent and hominid agents will show an exponentially increasing trend under the scenario without climate changes and a steadily growing trend under the scenario with climate changes. In summary, the agents will get more wealth under the scenario without climate changes in the long run.

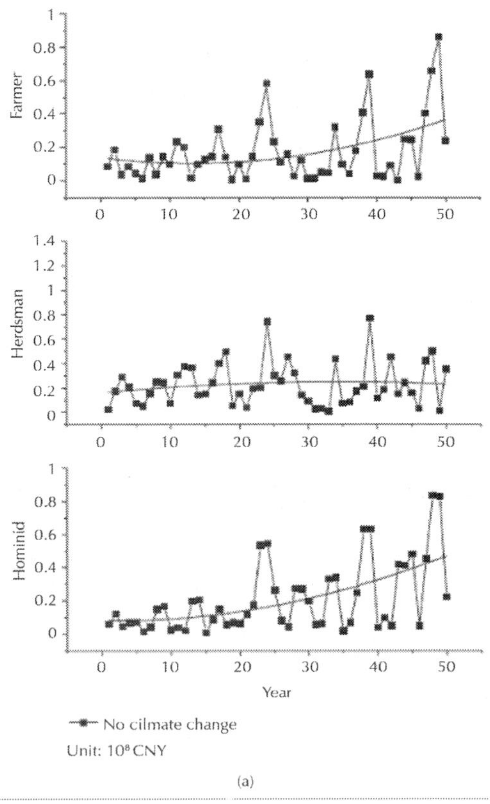

(a)

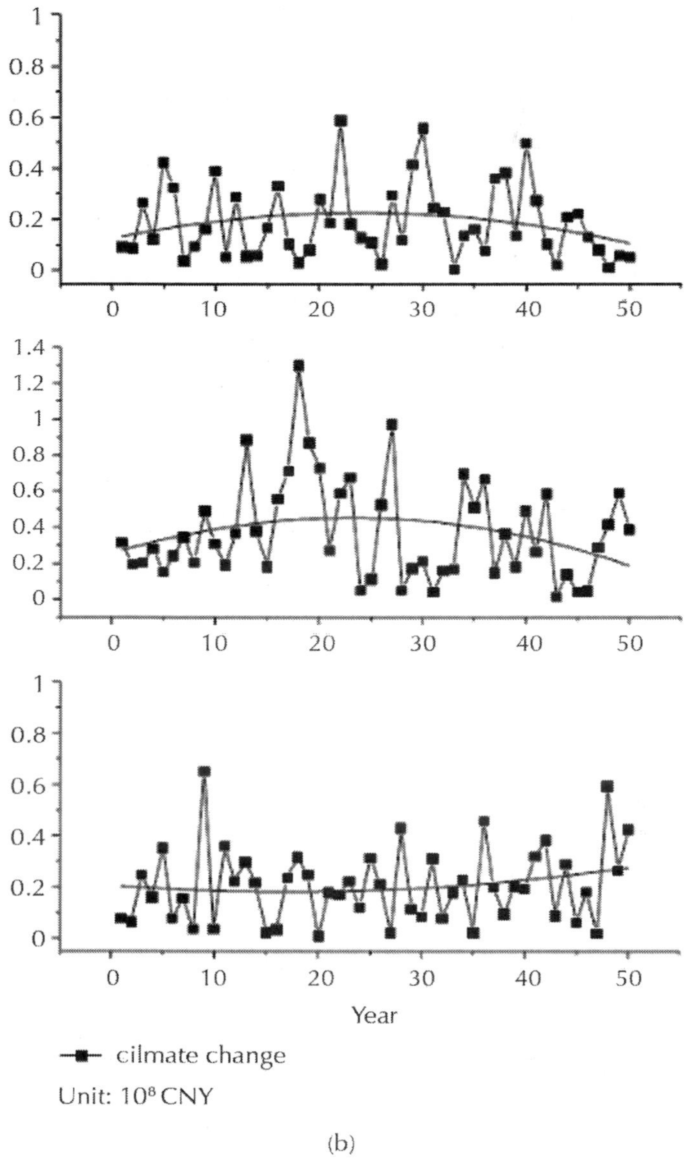

(b)

Figure 3: Simulation result of annual income change of agents under the scenario with (b) and without climate changes (a).

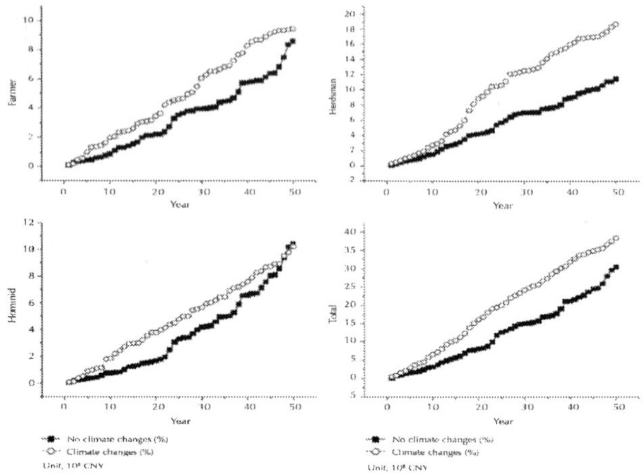

Figure 4: Simulation result of the change of total income of agents under different climate change scenarios.

Sensitivity Analysis

In order to analyze sensitivity of the cell numbers of cultivated land and grassland, total income of agents, and yield of land parcels to the driving factors listed in Table 1, the relative sensitivity was analyzed with the following formula:where is the simulated result after the change of inputs, is the background result, and is the change of input variables. The relative sensitivity represents the change of the result after the change of only one variable. The greater the change of the results is, the more sensitive to the input variables the results are. This study mainly focused on the sensitivity to the climate changes, market price, government policies, and cost control.

The result of sensitivity analysis indicated that the grid numbers of the cultivated land and grassland have a higher sensitivity to the prices of products of the agriculture and animal husbandry, government subsidies, and cost control (Table 2). Besides, the total income is more sensitive to the prices of products of the agriculture and animal husbandry, cost control, government subsidies, and precipitation. What is more, yield has a higher sensitivity to the cost control, precipitation, the prices of products of the agriculture and animal husbandry, and temperature. In summary, the prices of products of agriculture and animal husbandry, precipitation, subsidies from government for land use change, and cost control have significant influence on the land use changes and economic return in the study area.

Table 2: Results of the sensitivity analysis, unit: %

Variables	Cell numbers of cultivated land	Cell numbers of grassland	Total income	Yield
Temperature	1.05	−0.26	−0.16	1.31
Precipitation	−0.66	1.30	1.93	2.09
Price of products of agriculture	1.19	0.16	0.21	0.95
Price of products of animal Husbandry	−0.77	3.62	8.31	1.95
Government subsidies	−1.75	2.97	2.07	0.25
Cost control	−1.97	2.02	2.00	1.20

CONCLUSIONS

The climate changes can alter the land surface properties through influencing the human land use practices, and the ABM model is a good tool to study these climatic effects. The Three-River Headwaters Region was selected as the study area, where it is quite sensitive to the climate changes but has the characteristics of moderate land use changes and agents that are difficult to identify. The agents are identified and extracted and parameterized through analyzing the survey data and county level statistical data. Then the ABM model was established and used to simulate the dynamic land use changes under different climate change scenarios. In addition, CBR as an approach that is similar to the agents' thinking way in real life was used to parameterize the agents' behaviors in this study.

The results showed that the future trends of land use changes in the study area will be mainly characterized by the increase of grassland and decrease of unused land. A set of factors in terms of climate changes, agents' decisions, and government policies will affect the dynamic process of regional land surface. Besides, the result indicated that the model is quite sensitive to the climate changes, prices of agricultural and animal husbandry products, government subsidies, and cost control. Therefore, more importance should be attached to market price adjustment, stimulating policies implementation, and emerging technologies promotion, to lead reasonable land use practices so as to promote the reasonable land use practices, maintain the integrality and sustainability of ecosystems, and guarantee the provision of ecosystem services in the long term.

In this study, only the climate changes and agents' decisions on land use were taken into account. However, the household survey indicates that the agents' behaviors may be greatly affected by the governmental subsidies and the market prices that influence their economic income, which can be also illustrated with the sensitivity analysis. Therefore, more importance should be attached to the socioeconomic state when analyzing its impacts on the land use and terrestrial surface change in the future researches.

ACKNOWLEDGMENTS

This research was supported by the National Basic Research Program of China (973 Program) (no. 2012CB95570001), the National Key Programme for Developing Basic Science in China (no. 2010CB950904), and the Key Project in the National Science & Technology Pillar Program of China (no. 2013BAC03B00). Exploratory Forefront Project for the Strategic Science Plan in IGSNRR, CAS, is also greatly appreciated.

REFERENCES

1. H. O. Pörtner and A. P. Farrell, "Physiology and climate change," Science, vol. 322, no. 5902, pp. 690–692, 2008.
2. D. Schröter, W. Cramer, R. Leemans et al., "Ecosystem service supply and vulnerability to global change in Europe," Science, vol. 310, no. 5752, pp. 1333–1337, 2005.
3. R. B. Alley, J. Marotzke, W. D. Nordhaus et al., "Abrupt climate change," Science, vol. 299, no. 5615, pp. 2005–2010, 2003.

4. A. Jentsch and C. Beierkuhnlein, "Research frontiers in climate change: effects of extreme meteorological events on ecosystems," Comptes Rendus—Geoscience, vol. 340, no. 9-10, pp. 621–628, 2008.
5. S. Kröpelin, D. Verschuren, A. M. Lézine et al., "Climate-driven ecosystem succession in the Sahara: the past 6000 years," Science, vol. 320, no. 5877, pp. 765–768, 2008.
6. K. F. Drinkwater, G. Beaugrand, M. Kaeriyama et al., "On the processes linking climate to ecosystem changes," Journal of Marine Systems, vol. 79, no. 3-4, pp. 374–388, 2010.
7. J. Jin, S. Lu, S. Li, and N. L. Miller, "Impact of land use change on the local climate over the Tibetan Plateau," Advances in Meteorology, vol. 2010, Article ID 837480, 6 pages, 2010.
8. M. Wang, X. Zhang, and X. Yan, "Modeling the climatic effects of urbanization in the Beijing-Tianjin-Hebei metropolitan area," Theoretical and Applied Climatology, vol. 113, no. 3-4, pp. 377–385, 2013.View at Publisher · View at Google Scholar
9. S. Hitz and J. Smith, "Estimating global impacts from climate change," Global Environmental Change, vol. 14, no. 3, pp. 201–218, 2004.
10. T. Yue, Z. Fan, C. Chen, X. Sun, and B. Li, "Surface modelling of global terrestrial ecosystems under three climate change scenarios," Ecological Modelling, vol. 222, no. 14, pp. 2342–2361, 2011.
11. M. Ostwald and D. Chen, "Land-use change: impacts of climate variations and policies among small-scale farmers in the Loess Plateau, China," Land Use Policy, vol. 23, no. 4, pp. 361–371, 2006.
12. S. Briner, C. Elkin, R. Huber, and A. Grêt-Regamey, "Assessing the impacts of economic and climate changes on land-use in mountain regions: a spatial dynamic modeling approach," Agriculture, Ecosystems and Environment, vol. 149, pp. 50–63, 2012.
13. T. Li, R. F. Grant, and L. B. Flanagan, "Climate impact on net ecosystem productivity of a semi-arid natural grassland: modeling and measurement," Agricultural and Forest Meteorology, vol. 126, no. 1-2, pp. 99–116, 2004.

14. A. Fleischer and M. Sternberg, "The economic impact of global climate change on Mediterranean rangeland ecosystems: a space-for-time approach," Ecological Economics, vol. 59, no. 3, pp. 287–295, 2006.
15. L. E. Rustad, "The response of terrestrial ecosystems to global climate change: towards an integrated approach," Science of the Total Environment, vol. 404, no. 2-3, pp. 222–235, 2008.
16. C. E. Williamson, J. E. Saros, and D. W. Schindler, "Climate change: sentinels of change," Science, vol. 323, no. 5916, pp. 887–888, 2009.
17. M. D. A. Rounsevell and D. S. Reay, "Land use and climate change in the UK," Land Use Policy, vol. 26, supplement 1, pp. S160–S169, 2009.
18. L. Zhen, L. Yang, G. D. Xie, and S. K. Cheng, "Land use dynamics and policy implications in the Jinghe watershed of western China: a critical assessment from local perspectives," Natural Resources Journal, vol. 46, no. 4, pp. 859–879, 2006.
19. X. Xu, J. Liu, Q. Shao, and J. Fan, "The dynamic changes of ecosystem spatial pattern and structure in the three-river headwaters region in Qinghai Province during recent 30 years," Geographic Research, vol. 27, no. 4, pp. 829–839, 2008 (Chinese).
20. J. G. Polhill, L. Sutherland, and N. M. Gotts, "Using qualitative evidence to enhance an agent-based modelling system for studying land use change," Journal of Artificial Societies and Social Simulation, vol. 13, no. 2, p. 10, 2010.
21. L. R. Izquierdo, N. M. Gotts, and J. G. Polhill, "Case-based reasoning, social dilemmas, and a new equilibrium concept," Journal of Artificial Societies and Social Simulation, vol. 7, no. 3, p. 1, 2004.

Chapter 9

Critical Studies on Integrating Land-Use Induced Effects on Climate Regulation Services into Impact Assessment for Human Well-Being

Zhihui Li[1, 2, 3], Xiangzheng Deng[1, 3], Jikun Huang[1, 3], Rongrong Zhang[4], and Juan Huang[5] Yang[5]

[1]Institute of Geographic Sciences and Natural Resources Research, Chinese Academy of Sciences, Beijing 100101, China

[2]University of Chinese Academy of Sciences, Beijing 100049, China

[3]Center for Chinese Agricultural Policy, Chinese Academy of Sciences, Beijing 100101, China

[4]Faculty of Resources and Environmental Science, Hubei University, Wuhan 430062, China

[5]State Key Laboratory of Water Environment Simulation, School of Environment, Beijing Normal University, Beijing 100875, China

ABSTRACT

It is commonly acknowledged that land use changes (LUC) and climate changes have exerted significant effects on ecosystem services which are essential and vital to human well-being. Among all the services provided by ecosystem, climate regulation services are relatively sensitive to LUC and climate changes. This study aims to comprehensively review studies on the complex effects of LUC and climate changes on climate regulation services and further integrates the effects on climate regulation services into impact assessment for human well-being. In this study, we firstly introduced research efforts in which the drivers of and their corresponding effects on climate regulation services are briefly identified. Then, we explicitly reviewed the researches on the effects of LUC and climate changes on climate regulation services, especially focused on the certain methods and models used to quantify the effects on the major drivers of climate regulation services. After that, the effects of LUC and climate changes on human well-being via climate regulation services were revisited and commented accordingly. Finally, this paper discussed the current research gaps and proposed some research prospects in future studies.

INTRODUCTION

Land use changes (LUC) and climate changes are two major factors that result in the changes of ecosystem services [1–4]. Along with the socioeconomic development and emerging ecological environmental problems, global changes, and ecosystem services are becoming the research hot topics. The relationships among LUC, climate changes and ecosystem services are interlaced and complex, in which temporal and spatial variations in human-induced LUC and climate variability can result in the difference of ecosystem services [5].

Natural ecosystem delivers a lot of benefits to human beings, and these benefits are known as ecosystem services. According to the Millennium Ecosystem Assessment (2005), these ecosystem services include provisioning services such as provision of food, water, timber, fiber, and genetic resources; regulating services such as the regulation of climate, floods, disease, and water quality as well as waste treatment; cultural services such as recreation, aesthetic enjoyment, and spiritual fulfillment; and supporting services such as soil formation, pollination, and nutrient cycling [6]. Among all these services, supporting services and regulating services underpin the delivery of other service categories [7]. What is more, there often exist trade-offs between different services when humans make management choices, which can change the type, magnitude, and relative mix of services provided by ecosystems [8]. However, people generally prefer provisioning and cultural services over regulating services [9] thus tend to undervalue regulating services. Consequently, decision-makers often ignore these regulating services in ways that will seriously undermine the long-term existence of provisioning services [7]. The regulating services provided by ecosystems are diverse, among which climate regulation is a final ecosystem service. The ecosystems regulate climate through biogeochemical and biogeophysical processes [6], as sources or sinks of greenhouse gases (GHGs) and as sources of aerosols all of which affect temperature and cloud formation [10–13].

The processes involved in climate regulation include the following: (1) through the CO_2 in the atmosphere was absorbed photosynthesis; (2) evapotranspiration from soils and plants controls the amount of water vapour entering the atmosphere, thus regulating cloud formation and the radiative properties of the atmosphere; (3) the change of the albedo of different land surfaces can affect the climate; for example, the change in vegetation can have a cooling or heating effect on the surface climate and may affect precipitation; (4) the regulation of aerosols comes from soil erosion or vegetation through vegetation scavenging, which affects radiative heating of the atmosphere, surface albedo, and so forth [12].

There are many direct and indirect drivers that can affect the process of climate regulation services. For the systematic understanding of the effect mechanism and quantitative evaluation of the effects on climate regulation, it is important to clarify the major drivers and quantitatively analyze the effects induced by LUC and climate changes on those drivers through both biogeochemical and biophysical process. Comparatively, a direct driver more unequivocally influences ecosystem processes, while an indirect driver operates more diffusely by altering one or more direct drivers; that is, direct drivers have much more explicit effects on ecosystem processes [14] and usually cause physical change that can be identified and monitored [15]. As to climate regulation services, the indirect drivers mainly include demographic drivers (population growth and distribution, migration and ethnicity, etc.), economic drivers (economic growth and consumer choice, market force, industry size and globalization, etc.), and sociopolitical drivers (legislation, regulation, etc.) [16]. The direct drivers are listed in Table 1, among which land use drivers are the most important in the ecosystem context and can be identified as the main drivers of climate regulation [17], while over longer term, climate changes will also have feedback to climate regulation services [18].

Table 1: Direct drivers of changes in climate regulation and their corresponding effects, adapted from [18]

Category of drivers	Change of drivers	Effect on climate regulation services
Habitat change: land and sea use	Productive area: expansion, conversion, and abandonment (agriculture, forestry) Mineral and aggregate extraction (peat) Urbanization and artificial sealed surfaces	Affects carbon sinks and existing stores, greenhouse gas (GHG) emissions, albedo and evapotranspiration, shade, and shelter Affects soil carbon stores, GHG emissions Affects soil carbon stores, albedo, shade, shelter, local temperatures, and humidity
Pollution and nutrient enrichment	Pollution emissions and deposition Nutrient and chemical inputs	Affects aerosol sources (soot) Affects GHG emissions
Harvest levels/ resource consumption	Livestock stocking rates	Affects GHG emissions

Climate variability and change	Temperature and precipitation CO2 and ocean acidification Sea-level change	Affects existing carbon stores, evapotranspiration
		Affects existing carbon stores, GHG emissions, and aerosol sources
		Affects existing carbon stores

In this study, we focus on choosing close researches to explore how LUC and climate changes affect the climate regulation through biogeochemical and biogeophysical processes, respectively. Then incorporating with the effects on climate regulation services and researches on impact assessment for human well-being were further revisited. This study, firstly describes the climate regulation services and the needs and significance to study climate regulation services, then examines how LUC and climate changes affect climate regulation services through review of the researches on major drivers, and after that outlines how changes of the role of ecosystem services in regulating climate affect human well-being through investigation of four major aspects (economic value, extreme weather, food security, and human health) that are closely related with human well-being. This review study intends to provide a reference for the future research on LUC, climate changes, climate regulation services, and human well-being. The framework for the review about the effects induced by LUC and climate changes on climate regulation and further impact assessment for human well-being is shown in Figure 1.

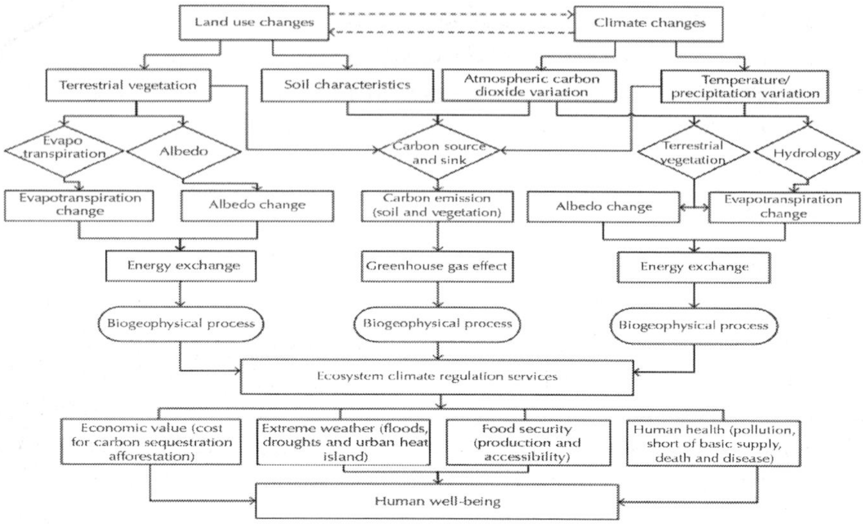

Figure 1: Framework for integrating land-use induced effects on climate regulation services into impact assessment for human well-being.

RESEARCH ON LUC INDUCED EFFECTS ON CLIMATE REGULATION SERVICES

As mentioned above, LUC plays an important role in climate regulation. Anthropogenic land use has been and will continue to be a major driver of the changes in climate system [17]. And there have been many observations and simulations revealing that LUC exerts effects on climate regulation services through biogeochemical and biogeophysical processes.

Consideration of the Cumulative Effects through Biogeochemical Processes

In terms of biogeochemical processes, LUC mainly affects the climate regulation services through the emission and sequestration of GHGs, especially through altering CO_2 flux. The total amount of carbon stored in terrestrial biosphere is an important factor in climate regulation [19]. Terrestrial ecosystems contribute to climate regulation primarily through carbon dynamics. Plants absorb CO_2 through photosynthesis, storing carbon in vegetation and soils, and the carbon accumulated in soil and biomass represents a pool of carbon which is greater than the atmospheric carbon pool [20]. Deforestation, forest degradation, and other land use practices accounted for approximately 20% of global anthropogenic CO_2 emissions during the 1990s [21]. When carbon is released from the earth during cultivation, deforestation, fire, and other land use practices, it binds with other chemicals to form GHGs in the atmosphere and accelerates global climate changes [21, 22]. The conservation of carbon sinks or pools is therefore important to mitigate GHGs levels. Thus it is of great significance to investigate the effect of LUC on climate regulation through its biogeochemical process, that is, through the effects on the cycle of carbon. And LUC can change the release of carbon to the atmosphere mainly through the disturbance on terrestrial vegetation and soils [23].

Biogeochemical Process Related with Terrestrial Vegetation

Terrestrial vegetation is a large carbon sink which plays an important role in the global carbon cycle and is valued globally for the services it provides to society. Vegetation classifications have been related to climate variables and used in assessment of possible global response to climate changes [24–26]. Olson et al. (1983) built up a computerized data base to document the map of vegetation and corresponding carbon density for natural and modified complexes of ecosystems. The map provides a basis for making improved estimates of vegetation areas and carbon quantities [27], illustrating that different types of vegetation have various ability of carbon sequestration. LUC can have great effects on the structure of terrestrial vegetation [28]; for example, once the forest is being converted to cultivated land, the biodiversity will decline. The distribution of sources and sinks of carbon among the world's terrestrial ecosystem is uncertain, through deforestation, urbanization, expansion of cultivated land, and other land use practices; LUC variously alter the land surface and species compositions and exert various effects on the carbon cycle.

It is relatively difficult to quantify the process that LUC affect the carbon sink or net source of terrestrial vegetation. Most researches developed and applied different models to record the carbon emission or sequestration resulted from LUC. Firstly, many researchers adopted empirical data to simulate carbon emission and sequestration. Houghton et al. (1983, 1999, 2000, and 2003) calculated the carbon emission resulting from LUC and the potential for sequestering carbon of different land covers mainly based on the parameters set for each vegetation type in each regime [29–36], and the model was called the "bookkeeping" terrestrial carbon model. Further, Houghton et al. (2001) documented a numeric data package that consists of annual estimates of the net flux of carbon between terrestrial ecosystems and the atmosphere resulting from deliberate LUC, especially forest clearing for agriculture and the harvest of wood for wood products or energy from 1850 through 1990 [37]. DeFries et al. (2002) further apply the "bookkeeping" model in conjunction with multiple sources of remote-sensing data to estimate carbon fluxes from tropical LUC [38]. The "bookkeeping" model tracks the amount of carbon released to the atmosphere from clearing and decay of plant material and pluses the amount of carbon accumulated as vegetation grows back. However, the parameters in the "bookkeeping" model are mainly based on empirical data and lack mechanism process. Then some researches started to combine process-based ecosystem models with the "bookkeeping" model, which can associate the land use information with its related process and thus enhanced the accuracy of the estimation. For example, one type of the process-based ecosystem model, the Terrestrial Ecosystem Model (TEM), in which the effect of LUC was characterized by the specific parameters of the function of different vegetation types, can be combined with the "bookkeeping" [19]. And DeFries et al. (1999) predicted the effect of LUC on carbon cycle based on the vegetation distribution information derived from remote sensing data and combined with Carnegie-Ames-Stanford Approach Model (CASA) [39]. What is more, Sitch et al. (2003) also coupled the "bookkeeping" model with Lund-Potsdam-Jean dynamic global vegetation model (LPJ) [40]. Particularly, McGuire et al. (2001) combined four models, that is, TEM, High Resolution Biosphere Model (HRBM), LPJ, and Integrated Biosphere Simulator (IBIS) model [19]. And Levy et al. (2004) combine the HYBRID vegetation dynamic model with the "bookkeeping" model to estimate the carbon dynamic resulted from LUC [41]. Differently, Schröter et al. (2005) used a range of ecosystem models and scenarios of climate and LUC to assess the vulnerability of ecosystem services, and the simulation results showed that LUC affected terrestrial carbon sink positively through decreases in agricultural land and increased afforestation during the 21th century [1]. Different models have different results about the amount of the carbon emission induced by LUC, but all the results show that global deforestation can lead to large carbon emission.

Among all the vegetation types, forest as one of the productive area is a major part of terrestrial ecosystem. Global deforestation has great effect on the carbon emission [42]. Using forest inventory data and long-term ecosystem carbon studies, Pan et al. (2011) estimated a total forest sink of 2.4 ± 0.4 petagrams of carbon per year (Pg C year^{-1}) globally for 1990–2007, and tropical deforestation around the world contributes to a gross emission of 2.9 ± 0.5 Pg C year^{-1}, which was equal to 40% of global fossil fuel emissions [43]. Combined with spatial explicit information on changes in forest area which derived from satellite observations and terrestrial carbon models, many researches focused on the tropics to detect whether tropic deforestation act as carbon sink or net source and identified that tropics acted as carbon sink [38, 44]. Explicitly, Arora et al. (2009) assessed the biogeochemical behavior of the CCCma earth system model (CanESM1) against observations for 1850–2000 and then compared simulated atmospheric CO_2 concentration with available observations and observation-based estimates; the simulation results showed that forests' different photosynthesis ability and CO_2 emissions from LUC resulted in different density of CO_2 in atmosphere, which indicated that the tropics were large carbon sinks [45].

Currently, based on model simulation, it was acknowledged that LUC can affect the carbon sink or source. Firstly, many researches were based on empirical data and statistical models which can use only observation data about the land cover without knowing conversions among different land cover types. While with the application of remote sensing and developed ecosystem models, the carbon emission induced by LUC can be more accurately estimated. However, most models also cannot take some important factors into account, such as the feedback of regional climate change, the water cycle before and after LUC, and the physical structure of soil. More efforts should be devoted to develop models that can comprehensively integrate all the factors to do simulations.

Biogeochemical Process Related with Soil Carbon

The changes of LUC can affect the sink of carbon not only through the sequestration of vegetation covering the land, but also through the changes of soil characteristic of different land cover types. It has been identified that different land cover types' potentials of soil carbon are various. For example, tropical forests were estimated to store approximately 206 Pg C in the soil globally [4], which was relatively less than half of that of boreal forests' soil [18, 46]. And explicitly, wet and moist tropical forests tend to have greater soil carbon pools per unit area than tropical dry forests due to higher rates of Net Primary Productivity (NPP) [47]. The potential of cropland soils [48] and grazing land soils [49] to sequester carbon and mitigate the greenhouse effect are different. Dawson and Smith (2009) reviewed effects of LUC and climate changes on soil carbon losses and found that the soil carbon emissions of agriculture land (cropland and grassland), forestry land, and peatlands/wetlands explicitly have different ability of carbon sequestration, and human's land use management is critical to regulate carbon emission [50].

LUC plays an important role in determining whether soil is a sink or source of atmospheric CO_2, since it can change the amount of soil carbon sequestration. For example, deforestation can result in an initial loss of carbon from soils due to increased decomposition rates, erosion, and reduced inputs of vegetation residues, namely, soil organic matters (SOM) [51]. Ayanaba et al. (1976) reported that deforestation for cultivated land leads to the reduction of soil total carbon content [52]. As most researches about the effect of deforestation on soil carbon only focused on one kind of land conversion while did not simultaneously take all the land cover types into consideration, Powers (2004) presented better understanding of the effects of LUC on regional and global ecosystem process by considering all of the land use conversions that occur in the landscape and detect the difference among the effects on soil carbon of each land use conversion based on statistical analyze [53]. Scott et al. (2002) designed a soil carbon monitoring system for New Zealand using country-specific land use and soil carbon information, which can be used to estimate both soil carbon stocks at a single point in time and also future soil carbon stocks in response to LUC [54].

LUC affect the soil carbon content mainly through regulating soil organic carbon (SOC) pools [55]. The dynamic changes of SOC have a strong effect on atmospheric composition and the rate of climate change, thus playing an important role in climate regulation. There have been many studies quantifying the effects of LUC on SOC, and since the carbon dynamics in soil are a long-term process which is relatively difficult for doing field researches and observations, most of the studies were based on model simulations. The major SOM models include SOM model (SOMM), Institute of Terrestrial Ecology-Edinburg (ITE), Verberne, Rothamsted C model (ROTHC), Carbon-Nitrogen-Dynamics (CANDY), DeNitrification and DeComposition (DNDC), CENTURY, DAISY, and NCSOIL, and so forth. Smith et al. (1997) compared these nine models with data from seven long-term experiments representing multiple vegetation and management conditions [56]; the results showed that a comparison of the overall performance of models across all datasets reveals that the model errors of one group of models (ROTHC, CANDY, DNDC, CENTURY, DAISY, and NCSOIL) did not differ significantly from each other. Another group (SOMM, ITE, and Verberne) did not differ significantly from each other but showed significantly larger model errors than did models in the first group. Among all the nine models, ROTHC and CENTURY are two of the most widely used and tested SOM models. Shrestha et al. (2009) used the Century model to simulate changes in SOC pool over 100 years (1950–2050) under managed dense Shorea forest, rainfed upland, and irrigated low land in a midhill watershed of Nepal [57]. And Dieye et al. (2012) analyzed the sensitivity of the GEMS soil organic carbon model in response to LUC [58]; the GEMS model was also developed from the CENTURY model [59]. Barancikova et al. (2010) and Francaviglia et al. (2012) used the ROTHC model to simulate the changes in agricultural land's SOC pools [60, 61]. SOC cannot only be affected by LUC, but also by climate change, which will be revisited in Section 3.1.

Combined Biogeochemical Process of Terrestrial Vegetation and Soil

As mentioned above, most researches focused on the calculation of carbon emission resulted from the terrestrial vegetation changes with different models and methodologies, based on the historical observations or scenarios analysis. And also there are researches trying to figure out the change of carbon emission as the change of soil characteristic of different land cover types resulted from land use management practices. However, it is also important to combine the two biogeochemical processes together to detect the effect of LUC on carbon cycle and further analyze the effect on climate regulation services.

As to this point of view, Woomer et al. (2004) defined the total carbon in the ecosystem as the sum of the woody biomass, herbaceous biomass, root, litter, and soil carbon pools, and using an inventory procedure involving satellite images which reveal historical LUC and recent field measurements of standing carbon stocks occurring in soil and plant; they estimated Senegal's terrestrial carbon stocks in 1965, 1985, and 2000 [62]. In addition, the changes of carbon emission from vegetation and soil resulted from LUC and climate changes can also counteract with each other. Synthetically considering the carbon emission from vegetation and soil induced by LUC and climate changes, based on a series of ecosystem models and scenarios of LUC and climate changes to evaluate the ecosystem service supply and vulnerability, Schröter et al. (2005) confirmed that Europe's terrestrial biosphere acted as a net carbon sink, with decreased agricultural land and increased afforestation resulting in the increasing sequestrated amount of carbon, while soil carbon losses due to warming would balance the carbon sequestrated by terrestrial biosphere by 2050 and lead to carbon releases by the end of this century [1]. In the Integrated Valuation of Ecosystem Services and Tradeoffs (InVEST) model, a simplified carbon cycle that maps and quantifies the amount of stored and sequestered carbon based on five carbon pools (above ground biomass, below ground biomass, soil, dead organic matter, and harvested wood products) is incorporated [63]. Leh et al. (2013) used average literature values in the InVest model to calculate the carbon as part of the ecosystem services in major land classes according to the proportion of land use land cover (LULC) area in Ghana and Cote d'Ivoire, West Africa [64]; the model is of great significance to evaluate the ecosystem services.

In summary, effect on climate regulation through biogeochemical process mainly focuses on the carbon emission and sequestration both in terrestrial vegetation and soil; from comprehensive and global perspective, it is important to regulate the carbon cycle to control global warming and there should be more researches to combine the whole biogeochemical processes of LUC's effect on carbon cycle together to provide references for decision-makers to make land use management practices.

Overall Effects Accounting through Biogeophysical Processes

LUC is a general term for the human modification of Earth's terrestrial surface, and the albedo and evapotranspiration of different surfaces vary with land cover types. The biogeophysical effect mechanisms of LUC on climate regulation are mainly through changing the albedo and evapotranspiration, which are closely related with energy flux between land surface and atmosphere [23, 65–70].

Biogeophysical Process Related with Albedo

Albedo represents the fraction of incoming radiation reflected by the surface. A reduction in albedo means that a larger fraction of the incoming radiative energy is absorbed by the surface, further results in warming [71]. The effect of albedo changes on regional and local climates is a research hotspot, especially changes in climate in response to changes of vegetated land cover and built-up areas. These changes alter surface heat balance not only by changing surface albedo, but also by altering evaporative heat transfer caused by evapotranspiration from vegetation and by changes in surface roughness [23].

Land surface with different vegetation types possesses different albedos; for example, forests typically have lower albedos than bare ground, grassland, and cropland; therefore, absorb more incoming solar radiation [12,17], indicating that different vegetation types participated in the land surface play a role in regulating surface energy, thus affecting surface climate. Based on different methodologies and models, there have been many researches focusing on the net radiation associated with the change of albedo resulted from LUC. In the 1970s, Charney (1975) firstly pointed out that albedo changes play an important role in global climate based on Global Climate Model (GCM) and put forward the biogeophysical feedback mechanism [72]. The mechanism explains that vegetation reduction due to drought will lead to increasing the albedo, decreasing the net radiation of the surface and corresponding sensible and latent heat, and further weakening the convergence upward movement, thus resulting in the reduction of cloud and precipitation. It can be seen that albedo plays an important role in surface net radiation and energy exchange. The radiative forcing due to surface albedo change can be simulated with the radiative transfer scheme of the third Hadley Centre Atmosphere Model (HadAM3) [73]. Betts (2000) applied HadAM3 to simulate the radiative forcing associated with changes in surface albedo and revealed that decreasing albedo exerts a positive radiative forcing on climate [65]. Snyder et al. (2004) investigated the participation of different vegetation types within the physical climate system based on CCM3-IBIS (the Community Climate Model coupled to the Integrated BIosphere Simulator), a coupled atmosphere-biosphere model, analyzed the effects of six different vegetation biomes (tropical, boreal, and temperate forests, savanna, grassland and steppe, and shrubland/tundra) on the climate through their role in modulating the biogeophysical exchanges of energy, water, and momentum between the land-surface and the atmosphere, and explained the role of the albedo in modifying the surface radiation budget [70].

Biogeophysical Process Related with Evapotranspiration

Evapotranspiration causes local cooling due to latent heat transfer from the surface to the atmosphere. Evapotranspiration can also influence cloud cover, which, in turn, affects the amount of energy reaching the surface [71]. Clearing vegetation reduces evapotranspiration and associated latent heat flux, as without the vegetation, energy normally used to evaporate water instead of heating the land [17]. Evapotranspiration is an important ecosystem integrity indicator and is strongly related to land cover types. The process of evapotranspiration consumes energy and therefore has a cooling effect and a positive effect on microclimate regulation. To quantify the climate regulation and assess the effect of LUC on climate regulation, Kroll et al. (2009) used thermal emissions of different land use classes with different land surface emissivity derived from remote sensing data and land cover information of MOLAND classification based on PLUREL scenario [74]. And Twine et al. (2004) analyzed the effects of LUC on the energy and water balance of the Mississippi River basin using the IBIS model, in which the forest cover was assumed to be completely converted to crop cover which resulted in decreasing annual average net radiation and evapotranspiration [75]. In order to assess the effect of LUC on evapotranspiration, Liu et al. (2008) used Dynamic Land Ecosystem Model (DLEM) in conjunction with spatial data of LUC to estimate the LUC's effects on the magnitude, spatial and temporal variations of evapotranspiration in China, and results showed that deforestation averagely increased evapotranspiration by 138 mm/year and urban sprawl generally decreased evapotranspiration 98 mm/year during 1900–2000, and so forth [76]. Yang et al. (2012) first used a knowledge-based decision tree (K-DT) classification technique to detect LUC which was characterized with deforestation and expansion of farmland, barren land and residential land, then a two-source potential evapotranspiration (PET) model was used to estimate the potential evapotranspiration response to LUC, and the result showed a decreasing trend of PET at Shalamulun River watershed in China [77]. And more efforts on quantifying the biogeophysical regulation of climate by ecosystems have largely focused on regional analyses, using GCMs that include land-surface models coupled with atmospheric circulation models, and these regional-scale analyses have focused on areas with strong land-climate feedbacks, including the Amazon [66, 78] and boreal regions [79].

As mentioned above, current approaches for quantifying the biogeophysical regulation of climate by ecosystems (especially through effects on atmospheric heat and moisture) were mainly concentrated on highly complex models, especially atmospheric models, such as HadAM3 and GCMs; thus few nonmeteorologists have access (or the expertise) to run and interpret high complex systems; then West et al. (2010) put forward an alternative approach for quantifying climate regulation by ecosystem; they developed a simple climate regulation index that could quickly produce approximations of biogeophysical regulation of climate by terrestrial ecosystems, in which the potential effects of LUC on biogeophysical climate regulation were estimated based on comparing a natural vegetation scenario to a bare ground scenario [69]. However, conversion of natural land to other land-cover types can have different results; thus the model has potential to be improved.

In some sense, researches of effect on climate regulation of LUC through biogeophysical process are of great significance and mainly focused on the net radiation, energy balance and so forth, which is related with albedo change and evapotranspiration rate. And most researchers simulate the effects with complex atmospheric models, biosphere models, and combined models, and the effect mechanism is relatively similar, while more attention should be paid to how to integrate and simplify the whole process to support most policy-makers, nonmeteorologists, and other related people decision making.

IDENTIFICATION ON THE EFFECTS ON CLIMATE REGULATION SERVICES DUE TO CLIMATE CHANGES

According to the forth report of Intergovernmental Panel on Climate change (IPCC AR4, 2007), there is more than 90% probability that human activities have affected the climate [21], mainly through two approaches: fossil fuel burning and land use/cover change. LUC cannot only directly exert effect on the drivers of climate regulation, but also indirectly through its impact on climate changes [80–82]. And climate changes (temperature and precipitation variation and CO_2 variation) are also likely to put many ecosystem services that humans derive from lands, waters, and so forth at risk. Over the long term, climate changes will feed back to climate regulation services [1] also through both biogeochemical and biogeophysical process.

Biogeochemical Process of the Effect on Climate Regulation Services

The amount of carbon in atmosphere is important to climate regulation services. Climate variations, such as the changes of temperature, precipitation, and atmospheric CO_2 content affect existing carbon stores through biogeochemical process. Friedlingstein et al. (2006) used eleven coupled climate-carbon cycle models with a common protocol to study the coupling between climate changes and the carbon cycle, through the feedback analysis; their results indicated that future climate changes will reduce the efficiency of the earth system to absorb the anthropogenic carbon perturbation [83]. And as to carbon accumulation, soil carbon is a major component of global carbon inventory which interact with atmospheric CO_2. Rising atmospheric CO_2concentrations will increase radiative forcing and is expected to increase soil temperatures and accelerate decomposition rates of SOC, which, in return, will increase CO_2 accumulation rate in the atmosphere [84–86].

Regarding the temperature changes, which are major elements affecting the photosynthesis [87] and decomposition of SOM [88]. Jenkinson et al. (1991) used the Rothamsted model for the turnover of organic matter in soil to calculate the amount of CO_2 that would be released from the world stock of SOM if temperatures increase as predicted and the annual return of plant debris to the soil being held constant. The results showed that if world temperatures rise by 0.03°C per year, the additional release of CO_2 from SOM over the next 60 years will be 61×10^{15} gC [89]. The temperature's effect on soil carbon losses is also confirmed by recent experimental and modeling studies [90–93] and lots of experimental studies overwhelmingly indicate that increased SOC decomposition resulting from higher temperature led to increased CO_2 emissions from soils [91, 94, 95]. In addition to carbon emission, soil carbon sequestration can be regulated by temperature change [93, 96].

As to moisture or precipitation conditions, in mature tropical forests, soil carbon pools tend to decrease exponentially as the ratio of temperature to precipitation increases, corresponding to a gradient from wet to dry forests [47]. Albani et al. (2006) used ecosystem demography (ED) model, through comparison between the patterns of variability in net ecosystem productivity (NEP) with the patterns of temperature and precipitation variability from 1948 to 2003 and revealed that the periods of carbon losses during the 1960s were caused by anomalously dry conditions, while the carbon losses that occurred in the late 1990s were due to a combination of reduced precipitation and anomalously warm temperatures in the Eastern US [97]. What is more, to better quantify climate changes' effect on soil carbon, Ise and Moorcroft (2006) used a mechanistic decomposition model in which the effects of temperature and moisture are multiplicative, to estimate the global-scale temperature and moisture dependencies of SOC decomposition, the results indicated that modeling of temperature and moisture dependencies of SOC decomposition in global-scale models should consider effects of scale which were not considered in other models [88]. And as mentioned in Section 2.1.3, SOC can also be affected by climate change. In many SOM models, climate data are important part as the input data into the models, and SOC changes in response to climate changes under different climate scenarios which derived from GCMs and other climate models were usually analyzed [60, 61, 98]. What is more, as the biogeochemical process of carbon sequestration is complex, it is difficult to take into account of all influencing factors; Zhan et al. (2012) used a panel data model and decomposition analyses to figure out the subtle effects of climatic, demographic, geographic, and economic factors, which revealed the importance of climatic factors in carbon sequestration [99].

Biogeophysical Process of the Effects on Climate Regulation

Human induced climate changes are also important factors affecting the climate regulation services. Along with the industrialization and other human activities, the global climate showed a warming trend, which in the long-term will return feedback to climate regulation services through biogeophysical process.

Effect on Evapotranspiration

Climate changes can exert effects on the hydrological cycle and alter the evapotranspiration that both with implication for ecosystem services and feedback to regional and global climate. Land evapotranspiration is a central process in the climate system and a nexus of the water, energy, and carbon cycles [100]. As evapotranspiration cannot be measured directly at the scale of climate observations and climate predictions, therefore most researches generally applied hydrological models to estimate the effect of climate changes on evapotranspiration [101]; among the models, different PET models are often used. To accurately detect the effect of climate changes on PET, more and more researchers turned over to analyze the sensitivity of these models to climate changes [102–105].

Effect on Albedo due to Changes of Terrestrial Vegetation

Climate changes can also exert effect on climate regulation via the change of vegetation which plays an important role in climate regulation through albedo and evapotranspiration change. The response of terrestrial vegetation to climate changes has long been a research hotspot; there are many studies about understanding and quantifying climate-induced vegetation change [106, 107].

Regarding temperature and precipitation changes, Overpeck et al. (1990) used the FORENA model to simulate the effects of climate changes on four forest types in eastern North America with three types of climate change experiments; the results indicated that climate changes would induce increases in forest and vegetation disturbance, which can significantly alter the total biomass and compositional response of forests to future warming [108]. As anthropogenic increases in the atmospheric concentration of CO_2 and other GHGs are predicted to cause a warming of the global climate by modifying radiative forcing [109], Peters and Darling (1985) had a widespread hypothesis that global warming resulted from greenhouse effect can diminish biological diversity and shift the terrestrial pattern [110]. Moreover, corresponding climate changes with enriched CO_2 may also alter the density of vegetation cover, thus modifying the physical characteristics of the land surface to provide climate feedback. In a vegetation/ecosystem modeling and analysis project (1995), three biogeochemical models (BIOME-BGC (BioGeochemistry Cycles), CENTURY, and TEM) and three biogeophysical models (BIOME2, Dynamic Global Phytogeography Model (DOLY), and Mapped Atmosphere-Plant Soil System (MAPSS)) were applied, respectively, to simulate the geographic distribution of vegetation types under doubled CO_2 and a range of climate scenarios (temperature and precipitation variations), and the results indicated that carbon concentration and temperature and precipitation variations can obviously change the distribution of vegetation types [111]. Betts et al. (1997) used a general circulation model iteratively coupled with an equilibrium vegetation model to quantify the effects of both physiological and structural vegetation feedbacks on a doubled-CO_2 climate, with the output vegetation structure mainly represented by Leaf Area Index (LAI). The results showed that changes in vegetation structure can significantly feed back to regional-scale climate [112]. The vegetation structure feeds back to climate mainly through two opposite effects; increased LAI tended to warm the land surface by lowering its albedo and cool the land surface by enhancing evaporation, while decreased LAI tended to cool the surface via increased albedo and warm the surface via reduced evaporation [79, 113]. To understand how vegetation growth responded to the climatic variations, Piao et al. (2006) took the LAI as an indicator to represent the vegetation activity and based on a mechanism terrestrial carbon model, to detect effect of current climate changes on vegetation growth in the northern hemisphere, and results showed that the vegetation change in different areas can be explained by temperature and precipitation, respectively [114].

Most of the researches focused on terrestrial vegetation ecosystem, while Lovelock and Kump (1994) suggested that if living organism participates in climate regulation in an active and responsive way, they do most probably as part of a tightly coupled system, which includes the biota, the atmosphere, the ocean, and the crustal rock. Within such system, the growth of organisms changes environmental conditions and environmental changes feed back to the growth. Thus they conducted qualitative analysis of the effects of temperature change on the feedbacks induced by changes in surface distribution of both marine algae and land plants, as to detect how the planetary area occupied by these two ecosystems varies with temperature [115]. There are many kinds of ecosystems on the earth, among which terrestrial ecosystem is much more connected with human beings, while it is also of great significance to do researches that combine other ecosystems with terrestrial ecosystem to comprehensively detect the impact mechanism of LUC and climate changes.

IMPACT ASSESSMENT FOR HUMAN WELL-BEING DUE TO LAND-USE INDUCED CLIMATE REGULATION SERVICES CHANGES

According to the Millennium Ecosystem Assessment (2005), human well-being is assumed to have multiple constituents, including the basic material for a good life (adequate livelihoods, sufficient nutrient food, shelters, access to goods, etc.); health (feeling well, access to clean air and water, etc.); security (secure resource access, personal safety, and security from disasters, etc.), and freedom of choice and action and so forth [6]. The services provided by ecosystems for all those unalterable needs indicate that ecosystem services are essential to human well-being. Climate regulation relates to the maintenance of a favorable climate, both at local and global scales, which has important implications for health, crop productivity, and other human activities [7]. In this paper, we reviewed how changes in the role of ecosystem services in regulating climate affect human well-being, mainly including economy benefits, food security, human health, and a healthy environment.

Economic Value Determined by Climate Regulation

Ecosystem services and the natural capital stocks that produce the services are critical to support the earth's life system, contribute to human well-being both directly and indirectly, and therefore represent part of the total economic value of the planet. Costanza et al. (1997) firstly developed the Ecosystem Value System (ESV) calculation system to estimate the economic value of 17 ecosystem services (including climate regulation services) for 16 biomes, in which land use is one of the major factors [116]. Based on the ESV calculation system, some researchers examine the potential effects of the past and future LUC on the ecosystem services [117, 118], and the results showed that the changes of ecosystem services value (among which climate regulation accounted for 10% of the change) in Baguio city about 98% were due to forest cover loss for 1988–2009.

Forest ecosystems play an important role in climate regulation through trapping moisture and cooling the earth's surface, thus regulating precipitation and temperature. Costanza et al. (1997) found that forests yield US$450 per hectare per year in terms of climate regulation benefits [116]. Explicitly, in the urban context which is directly related with human well-being, urban and community forests can strongly influence the physical/biological environment and mitigate many effects of urban development by moderating climate, conserving energy, CO_2, and water, improving air quality, controlling rainfall runoff, and so forth. As in the case of Tucson, Arizona, each tree would give benefits in the range of US$20.75 annually by reducing cooling costs for buildings [119]. Dwyer et al. (1992) calculated that 100 million mature trees in US cities could reduce annual energy costs by US$2 billion based on computer simulations estimate [120].

LUC and climate changes can exert effect on the capacity of ecosystem to absorb carbon, which will result in the loss of climate regulation services and further loss of socioeconomic benefits. To estimate the value of carbon sequestration in Uganda's protected area, Howard (1995) used two different approaches: first, based on figures of the damage that would occur if the land was converted and carbon released in the atmosphere, the value of Uganda's protected areas as a carbon sink was estimated at US$245 million, which amounts to a US$17.4 million annually; second, the replacement cost approach was also used to estimate the cost of replacing this carbon sink functions through an afforestation scheme and was valued at US$20.3 million annually [121]. The results indicated that along with the degradation or loss of climate regulation services in carbon sequestration, much cost will be input to ecological engineering, such as afforestation and construction of environmental protection area.

Effects of Extreme Weather Events

In response to ecosystem services changes, extreme weather events directly affect human well-beings, leading to drastic disasters which can influence most aspects of humans, such as living environment, agriculture production, and health. Regional climatic conditions are influenced by changes in ecosystems and landscapes, especially deforestation and desertification. Human-induced alteration of atmospheric composition (the greenhouse effect) also affects climatic conditions. On a large scale, the long-term ongoing GHGs effect will lead to increasing risks of deaths in extreme weather events (heat waves, floods, droughts, etc.) [122]. These events have local and sometimes regional effects, directly through deaths and injuries and indirectly through economic disruption, infrastructure damage, and population displacement. In turn, this may lead to increased incidence of certain communicable diseases as a result of overcrowding, lack of clean water and shelter, poor nutritional status, and adverse effects on mental health.

Climate regulation of hydrology is a research area, in which natural climate system affect hydrology and further exert effect on the runoff, rainfall cyclicity, water quality, soil development, and so forth [123, 124]. As a result, there are marked variations in the likelihood of flooding and droughts, the intensity of erosion and nutrient cycling, and the demand for irrigation water and water supplies, which further will bring substantial loss of human well-being. And as to urban ecosystem, climate changes caused by increased anthropogenic emissions of CO_2 and other GHGs are a long-term climate hazard with the potential to alter the intensity, temporal pattern, and spatial extent of the urban heat island in metropolitan regions [125]. The microclimate regulation of all natural ecosystem in the urban areas contribute to reducing the urban heat island effect, as urban green increases evapotranspiration and therefore has a cooling effect; from the perspective of economics, vegetation can also decrease energy use for heating and air conditioning substantially in urban areas by shading house in summer and reducing wind speed in winter [126].

Response of Food Security to Climate Regulation

The accessibility of food is highly dependent on suitable climatic conditions, and the sustainable production of food and raw materials is vulnerable to changes in temperature, precipitation, and concentrations of CO_2. Though the role of climate regulation in food security has not been fully estimated [127], the climate variability which can result from LUC, has been examined as the principal source of fluctuations in global food production in different areas, such as the arid and semiarid tropical countries of the developing world, the northern China, and so forth [128–130]. Specifically speaking, soil carbon sequestration can be regulated by climate changes, especially the changes result from temperature change [93, 96] and LUC. As land is the basic requirement for cultivation of crops, a region's potential for food production in agricultural areas depends directly on the fertility of available arable lands [131], and soil carbon sequestration is a key element for ecosystem biogeochemical carbon cycle and contributes to the formation of fertility in soil, which helps to maintain net productivity, improve water quality, and restore degraded soils and ecosystems [20,

132]. Thus along with the climate regulation services changes, the variability of soil carbon sequestration results in the change of food production and accessibility.

Consequence on Human Health

In the Millennium Ecosystem Assessment, it was stated that stresses on freshwater sources, food-producing systems, and climate regulation could cause major adverse health effects, and each ecosystem service is sensitive to climate conditions which will be affected by LUC and climate changes. And the change of ecosystem services will affect the human well-being and health of human [122]. Climate regulation is one of the services the ecosystems provide to regulate climate conditions, mainly to ensure that people live an environmentally clean and safe life [133]. The degradation of climate regulation will decrease the ability of ecosystem to avoid the adverse effect of climate changes and can affect human health through both direct effect (such as increasing mortality, diseases from extreme weather) and indirect effects (such as climate-induced changes in the distribution of productive ecosystems and the availability of food, water, and energy supplies). Although climate changes may have some benefits to human health, most are expected to be negative especially in urban areas. As human migrate from rural areas to cities, more than half of the world's population now live in high-density urban areas, many of which are poorly supplied with either ecosystem or human services [122]. And in these urban areas, a reduction in climate regulation services could exacerbate climate stress for large numbers of people, reducing well-being and increasing death rates directly through higher summer temperatures and so forth [16]. Furthermore, climate regulation services of soil carbon sequestration can help mitigate climate changes by offsetting emissions of fossil fuels and reducing GHGs effects, and so forth, which is of great significance to improve environment for healthy lives [132, 134].

CONCLUDING REMARKS

LUC and climate changes will alter the supply of ecosystem services that are vital for human well-being. The processes through LUC and climate changes influence human well-being via exerting effects on climate regulation services are complex and complicated, involving both biogeochemical and biogeophysical processes. From the perspective of the drivers of and their corresponding effects on climate regulation services, this paper mainly revisited the closely related researches to explore how LUC and climate changes exert effects on the major drivers, such as carbon emission and sequestration, albedo, and evapotranspiration which are key elements to regulate surface energy exchange and further play an important role in climate regulation services. And, along with the changes of climate regulation services resulted from LUC and climate change, different aspects of human well-beings will be impacted which have also been assessed by many researchers. Generally speaking, LUC and climate changes have great effects on climate regulation services. Globally, deforestation, degradation of forest and grassland, land conversion from forest to cultivated land, urban expansion, carbon emission, and so forth all will probably lead to changes in climate regulation services.

So far, lots of researches have been conducted and significant progresses have been made related to each aspect shown in Figure 1. However, there is still room for improvement. First, there have been many methodologies and models developed or enhanced to investigate the effect mechanism of the drivers of climate regulation services via biogeochemical and biogeophysical process, while there are less researches that combine both the processes in the models; thus more research attention should be paid to the effect mechanisms that take both biogeochemical and biogeophysical effects into considerations. Second, in future research, improved models which can combine the effects of LUC and climate changes on the climate regulation as a whole should be developed. In addition, as human well-being changes along with the variation of climate regulation resulted from the effect of LUC and climate changes, further anthropogenic activities and economic development related with human well-being will respond to LUC and climate changes. However, there are relatively less researches that have taken the responses into consideration when assessing the effects, so future research should also focus on these potential societal responses.

ACKNOWLEDGMENTS

This research was supported by China National Natural Science Funds for Distinguished Young Scholar (Grant no. 71225005), National Key Program for Developing Basic Science in China (Grant no. 2012CB955700), and Key Projects in the National Science & Technology Pillar Program (Grant no. 2013BAC03B03).

REFERENCES

1. D. Schröter, W. Cramer, R. Leemans et al., "Ecology: ecosystem service supply and vulnerability to global change in Europe," Science, vol. 310, no. 5752, pp. 1333–1337, 2005.
2. M. J. Metzger, M. D. A. Rounsevell, L. Acosta-Michlik, R. Leemans, and D. Schröter, "The vulnerability of ecosystem services to land use change," Agriculture, Ecosystems & Environment, vol. 114, no. 1, pp. 69–85, 2006.
3. R. F. Bangash, A. Passuello, M. Sanchez-Canales et al., "Ecosystem services in Mediterranean river basin: climate change impact on water provisioning and erosion control," Science of the Total Environment, vol. 458, pp. 246–255, 2013.
4. S. N. Gosling, "The likelihood and potential impact of future change in the large-scale climate-earth system on ecosystem services," Environmental Science & Policy, vol. 27, pp. S15–S31, 2013.
5. X. Chen, J. Bai, X. Y. Li, G. P. Luo, J. L. Li, and B. L. Li, "Changes in land use/land cover and ecosystem services in Central Asia during 1990–2009," Current Opinion in Environmental Sustainability, vol. 5, no. 1, pp. 116–127, 2013.
6. W. Reid, H. Mooney, A. Cropper et al., Ecosystems and Human Well-Being: Synthesis, Millennium Ecosystem Assessment, Island Press, Washington, DC, USA, 2005.
7. P. Kumar, M. Verma, M. D. Wood, and D. Negandhi, Guidance Manual For the Valuation of Regulating Services, United Nations Environment Programme, 2010.

8. J. P. Rodríguez, T. D. Beard Jr., E. M. Bennett et al., "Trade-offs across space, time, and ecosystem services," Ecology and Society, vol. 11, no. 1, article 28, 2006.
9. S. R. Carpenter, E. M. Bennett, and G. D. Peterson, "Scenarios for ecosystem services: an overview,"Ecology and Society, vol. 11, no. 1, article 29, 2006.
10. W. J. Burroughs, Climate Change: A Multidisciplinary Approach, Cambridge University Press, Cambridge, UK, 2001.
11. J. Houghton, Global Warming: The Complete Briefing, Cambridge University Press, Cambridge, UK, 2004.
12. G. Bonan, Ecological Climatology: Concepts and Applications, Cambridge University Press, Cambridge, UK, 2nd edition, 2008.
13. D. Fowler, K. Pilegaard, M. A. Sutton et al., "Atmospheric composition change: ecosystems—atmosphere interactions," Atmospheric Environment, vol. 43, no. 33, pp. 5193–5267, 2009.
14. G. C. Nelson, E. Bennett, A. A. Berhe et al., "Drivers of change in ecosystem condition and services," inEcosystems and Human Well-Being: Scenarios:Findings of the Scenarios Working Group, Island Press, Washington, DC, USA, 2005.
15. N. Ash, N. Lucas, P. Bubb et al., "Framing the link between development and ecosystem services," inEcosystem Services: A Guide for Decision Makers, World Resouces Institute, Washington, DC, USA, 2008.
16. W. Jonathan, T. Megan, H. A. Louise et al., "The drivers of change in UK ecosystems and ecosystem services," UK National Ecosystem Assessment Technical Report 2011, UNEP-WCMC, Cambridge, UK.
17. K. J. Anderson-Teixeira, P. K. Snyder, T. E. Twine, S. V. Cuadra, M. H. Costa, and E. H. Delucia, "Climate-regulation services of natural and agricultural ecoregions of the Americas," Nature Climate Change, vol. 2, no. 3, pp. 177–181, 2012.
18. P. Smith, A. Mike, H. Black et al., "Regulating services," UK National Ecosystem Assessment Technical Report 2011, UNEP-WCMC, Cambridge, UK, 2011.
19. A. D. McGuire, S. Sitch, J. S. Clein et al., "Carbon balance of the terrestrial biosphere in the twentieth century: analyses of CO_2, climate and land use effects with four process-based ecosytem

models," Global Biogeochemical Cycles, vol. 15, no. 1, pp. 183–206, 2001.

20. R. Lal, "Soil carbon sequestration impacts on global climate change and food security," Science, vol. 304, no. 5677, pp. 1623–1627, 2004.
21. S. Solomon, D. H. Qin, M. Manning et al., Climate Change 2007: The Physical Science Basis, Cambridge University Press, Cambridge, UK, 2007.
22. R. Lal, "Soil carbon sequestration to mitigate climate change," Geoderma, vol. 123, no. 1-2, pp. 1–22, 2004.
23. E. Ellis and R. Pontius, Land-Use and Land-Cover Change, Encyclopedia of Earth, Environmental Information Coalition, National Council for Science and the Environment, Washington, DC, USA, 2007, http://www.eoearth.org/view/article/154143/.
24. W. R. Emanuel, H. H. Shugart, and M. P. Stevenson, "Climatic change and the broad-scale distribution of terrestrial ecosystem complexes," Climatic Change, vol. 7, no. 1, pp. 29–43, 1985.
25. K. C. Prentice and I. Y. Fung, "The sensitivity of terrestrial carbon storage to climate change," Nature, vol. 346, no. 6279, pp. 48–51, 1990.
26. T. M. Smith, R. Leemans, and H. H. Shugart, "Sensitivity of terrestrial carbon storage to CO_2-induced climate change: comparison of four scenarios based on general circulation models," Climatic Change, vol. 21, no. 4, pp. 367–384, 1992.
27. J. S. Olson, J. A. Watts, and L. J. Allison, Carbon in Live Vegetation of Major World Ecosystems, Oak Ridge National Lab, Oak Ridge, Tenn, USA, 1983.
28. P. Reidsma, T. Tekelenburg, M. van den Berg, and R. Alkemade, "Impacts of land-use change on biodiversity: an assessment of agricultural biodiversity in the European Union," Agriculture, Ecosystems & Environment, vol. 114, no. 1, pp. 86–102, 2006.
29. R. A. Houghton, J. E. Hobbie, J. M. Melillo et al., "Changes in the carbon content of terrestrial biota and soils between 1860 and 1980—a net release of CO_2 to the atmosphere," Ecological Monographs, vol. 53, no. 3, pp. 235–262, 1983.
30. R. A. Houghton and C. L. Goodale, "Effects of land-use change on the carbon balance of terrestrial ecosystems," Ecosystems and Land Use Change, vol. 153, pp. 85–98, 2004.

31. R. A. Houghton, "Revised estimates of the annual net flux of carbon to the atmosphere from changes in land use and land management 1850–2000," Tellus, Series B, vol. 55, no. 2, pp. 378–390, 2003.
32. R. A. Houghton and J. L. Hackler, "Changes in terrestrial carbon storage in the United States. I: the roles of agriculture and forestry," Global Ecology and Biogeography, vol. 9, no. 2, pp. 125–144, 2000.
33. R. A. Houghton and J. L. Hackler, "Sources and sinks of carbon from land-use change in China," Global Biogeochemical Cycles, vol. 17, no. 2, pp. 3–1, 2003.
34. R. A. Houghton, J. L. Hackler, and K. T. Lawrence, "The U.S. carbon budget: contributions from land-use change," Science, vol. 285, no. 5427, pp. 574–578, 1999.
35. R. A. Houghton, "The annual net flux of carbon to the atmosphere from changes in land use 1850-1990," Tellus, Series B, vol. 51, no. 2, pp. 298–313, 1999.
36. R. A. Houghton, J. D. Unruh, and P. A. Lefebvre, "Current land cover in the tropics and its potential for sequestering carbon," Global Biogeochemical Cycles, vol. 7, no. 2, pp. 305–320, 1993.
37. R. A. Houghton, J. L. Hackler, and R. M. Cushman, Carbon Flux to the Atmosphere from Land-Use Changes: 1850 to 1990, Carbon Dioxide Information Center, Environmental Sciences Division, Oak Ridge National Laboratory, 2001.
38. R. S. DeFries, R. A. Houghton, M. C. Hansen, C. B. Field, D. Skole, and J. Townshend, "Carbon emissions from tropical deforestation and regrowth based on satellite observations for the 1980s and 1990s," Proceedings of the National Academy of Sciences of the United States of America, vol. 99, no. 22, pp. 14256–14261, 2002.
39. R. S. DeFries, C. B. Field, I. Fung, G. J. Collatz, and L. Bounoua, "Combining satellite data and biogeochemical models to estimate global effects of human-induced land cover change on carbon emissions and primary productivity," Global Biogeochemical Cycles, vol. 13, no. 3, pp. 803–815, 1999.
40. S. Sitch, B. Smith, I. C. Prentice et al., "Evaluation of ecosystem dynamics, plant geography and terrestrial carbon cycling in the

LPJ dynamic global vegetation model," Global Change Biology, vol. 9, no. 2, pp. 161–185, 2003.

41. P. E. Levy, A. D. Friend, A. White, and M. G. R. Cannell, "The influence of land use change on global-scale fluxes of carbon from terrestrial ecosystems," Climatic Change, vol. 67, no. 2-3, pp. 185–209, 2004.
42. G. M. Woodwell, J. E. Hobbie, R. A. Houghton et al., "Global deforestation: contribution to atmospheric carbon dioxide," Science, vol. 222, no. 4628, pp. 1081–1086, 1983.
43. Y. Pan, R. A. Birdsey, J. Fang et al., "A large and persistent carbon sink in the world's forests," Science, vol. 333, no. 6045, pp. 988–993, 2011.
44. R. A. Houghton, D. L. Skole, C. A. Nobre, J. L. Hackler, K. T. Lawrence, and W. H. Chomentowski, "Annual fluxes of carbon from deforestation and regrowth in the Brazilian Amazon," Nature, vol. 403, no. 6767, pp. 301–304, 2000.
45. V. K. Arora, G. J. Boer, J. R. Christian et al., "The Effect of terrestrial photosynthesis down regulation on the twentieth-century carbon budget simulated with the CCCma earth system model," Journal of Climate, vol. 22, no. 22, pp. 6066–6088, 2009.
46. R. K. Dixon, S. Brown, R. A. Houghton, A. M. Solomon, M. C. Trexler, and J. Wisniewski, "Carbon pools and flux of global forest ecosystems," Science, vol. 263, no. 5144, pp. 185–190, 1994.
47. S. Brown and A. E. Lugo, "The storage and production of organic matter in tropical forests and their role in the global carbon cycle," Biotropica, vol. 14, no. 3, pp. 161–187, 1982.
48. J. M. Kimble, R. F. Follett, C. V. Cole, and R. Lal, The Potential of US Cropland to Sequester Carbon and Mitigate the Greenhouse Effect, Ann Arbor Press, Chelsea, Mich, USA, 1998.
49. R. F. Follett, J. M. Kimble, and R. Lal, The Potential of U.S. Grazing Lands to Sequester Carbon and Mitigate the Greenhouse Effect, CRC Press, New York, NY, USA, 2001.
50. J. J. C. Dawson and P. Smith, "Carbon losses from soil and its consequences for land-use management,"Science of the Total Environment, vol. 382, no. 2-3, pp. 165–190, 2007.
51. A. E. Lugo and S. Brown, "Management of tropical soils as sinks or sources of atmospheric carbon,"Plant and Soil, vol. 149, no. 1, pp. 27–41, 1993.

52. A. Ayanaba, S. B. Tuckwell, and D. S. Jenkinson, "The effects of clearing and cropping on the organic reserves and biomass of tropical forest soils," Soil Biology and Biochemistry, vol. 8, no. 6, pp. 519–525, 1976.
53. J. S. Powers, "Changes in soil carbon and nitrogen after contrasting land-use transitions in northeastern Costa Rica," Ecosystems, vol. 7, no. 2, pp. 134–146, 2004.
54. N. A. Scott, K. R. Tate, D. J. Giltrap et al., "Monitoring land-use change effects on soil carbon in New Zealand: quantifying baseline soil carbon stocks," Environmental Pollution, vol. 116, no. 1, pp. S167–S186, 2002.
55. R. Lal, "World soils and the greenhouse effect," IGBP Global Change Newsletter, vol. 37, pp. 4–5, 1999.
56. P. Smith, J. U. Smith, D. S. Powlson et al., "A comparison of the performance of nine soil organic matter models using datasets from seven long-term experiments," Geoderma, vol. 81, no. 1-2, pp. 153–225, 1997.
57. B. M. Shrestha, S. Williams, M. Easter, K. Paustian, and B. R. Singh, "Modeling soil organic carbon stocks and changes in a Nepalese watershed," Agriculture, Ecosystems & Environment, vol. 132, no. 1-2, pp. 91–97, 2009.
58. A. M. Dieye, D. P. Roy, N. P. Hanan, S. Liu, M. Hansen, and A. Toure, "Sensitivity analysis of the GEMS soil organic carbon model to land cover land use classification uncertainties under different climate scenarios in senegal," Biogeosciences, vol. 9, no. 2, pp. 631–648, 2012.
59. A. K. Metherell, Century: Soil Organic Matter Model Environment: Technical Documentation: Agroecosystem Version 4.0, Colorado State University, 1993.
60. G. Baran kov, J. Halas, M. Guttekova et al., "Application of RothC model to predict soil organic carbon stock on agricultural soils of Slovakia," Soil and Water Research, vol. 5, no. 1, pp. 1–9, 2010.
61. R. Francaviglia, K. Coleman, A. P. Whitmore et al., "Changes in soil organic carbon and climate change—application of the RothC model in agro-silvo-pastoral Mediterranean systems," Agricultural Systems, vol. 112, pp. 48–54, 2012.

62. P. L. Woomer, L. L. Tieszen, G. Tappan, A. Touré, and M. Sall, "Land use change and terrestrial carbon stocks in Senegal," Journal of Arid Environments, vol. 59, no. 3, pp. 625–642, 2004.
63. P. L. Woomer, A. Touré, and M. Sall, "Carbon stocks in Senegal›s Sahel transition zone," Journal of Arid Environments, vol. 59, no. 3, pp. 499–510, 2004.
64. M. D. K. Leh, M. D. Matlock, E. C. Cummings, and L. L. Nalley, "Quantifying and mapping multiple ecosystem services change in West Africa," Agriculture Ecosystems & Environment, vol. 165, pp. 6–18, 2013.
65. R. A. Betts, "Offset of the potential carbon sink from boreal forestation by decreases in surface albedo,"Nature, vol. 408, no. 6809, pp. 187–190, 2000.
66. M. H. Costa and J. A. Foley, "Combined effects of deforestation and doubled atmospheric CO_2concentrations on the climate of Amazonia," Journal of Climate, vol. 13, no. 1, pp. 18–34, 2000.
67. J. J. Feddema, K. W. Oleson, G. B. Bonan et al., "The importance of land-cover change in simulating future climates," Science, vol. 310, no. 5754, pp. 1674–1678, 2005.
68. R. A. Pielke Sr., G. Marland, R. A. Betts et al., "The influence of land-use change and landscape dynamics on the climate system: relevance to climate-change policy beyond the radiative effect of greenhouse gases," Philosophical Transactions of the Royal Society A, vol. 360, no. 1797, pp. 1705–1719, 2002.
69. P. C. West, G. T. Narisma, C. C. Barford, C. J. Kucharik, and J. A. Foley, "An alternative approach for quantifying climate regulation by ecosystems," Frontiers in Ecology and the Environment, vol. 9, no. 2, pp. 126–133, 2011.
70. P. K. Snyder, C. Delire, and J. A. Foley, "Evaluating the influence of different vegetation biomes on the global climate," Climate Dynamics, vol. 23, no. 3-4, pp. 279–302, 2004.
71. M. Carlson, J. Chen, S. Elgie et al., "Maintaining the role of Canada's forests and peatlands in climate regulation," Forestry Chronicle, vol. 86, no. 4, pp. 434–443, 2010.
72. J. G. Charney, "Dynamics of deserts and drought in Sahel," Quarterly Journal of the Royal Meteorological Society, vol. 101, no. 428, pp. 193–202, 1975.

73. V. D. Pope, M. L. Gallani, P. R. Rowntree, and R. A. Stratton, "The impact of new physical parametrizations in the Hadley Centre climate model: HadAM3," Climate Dynamics, vol. 16, no. 2-3, pp. 123–146, 2000.
74. F. Kroll, F. Müller, S. Bell et al., "Peri-urbanland use relationships—strategies and sustainability assessment tools for urban-rural linkages," integreted project, 2009.
75. T. E. Twine, C. J. Kucharik, and J. A. Foley, "Effects of land cover change on the energy and water balance of the Mississippi River basin," Journal of Hydrometeorology, vol. 5, no. 4, pp. 640–655, 2004.
76. M. Liu, H. Tian, G. Chen, W. Ren, C. Zhang, and J. Liu, "Effects of land-use and land-cover change on evapotranspiration and water yield in China during 1900-2000," Journal of the American Water Resources Association, vol. 44, no. 5, pp. 1193–1207, 2008.
77. X. Yang, L. Ren, V. P. Singh et al., "Impacts of land use and land cover changes on evapotranspiration and runoff at Shalamulun River watershed, China," Hydrology Research, vol. 43, no. 1-2, pp. 23–37, 2012.
78. J. Shukla, C. Nobre, and P. Sellers, "Amazon deforestation and climate change," Science, vol. 247, no. 4948, pp. 1322–1325, 1991.
79. G. B. Bonan, D. Pollard, and S. L. Thompson, "Effects of boreal forest vegetation on global climate,"Nature, vol. 359, no. 6397, pp. 716–718, 1992.
80. X. Z. Deng, C. H. Zhao, and H. M. Yan, "Systematic modeling of impacts of land use and land cover changes on regional climate: a review," Advances in Meteorology, vol. 2013, Article ID 317678, 11 pages, 2013.
81. E. J. Ma, A. P. Liu, X. Li, F. Wu, and J. Y. Zhan, "Impacts of vegetation change on the regional surface climate: a scenario-based analysis of afforestation in Jiangxi Province, China," Advances in Meteorology, vol. 2013, Article ID 796163, 8 pages, 2013.
82. R. J. Qu, X. L. Cui, H. M. Yan, E. J. Ma, and J. Y. Zhan, "Impacts of land cover change on the near-surface temperature in the North China Plain," Advances in Meteorology, vol. 2013, Article ID 409302, 12, pages, 2013.

83. P. Friedlingstein, P. Cox, R. Betts et al., "Climate-carbon cycle feedback analysis: results from the C4MIP model intercomparison," Journal of Climate, vol. 19, no. 14, pp. 3337–3353, 2006.

84. D. S. Schimel, B. H. Braswell, E. A. Holland et al., "Climatic, edaphic, and biotic controls over storage and turnover of carbon in soils," Global Biogeochemical Cycles, vol. 8, no. 3, pp. 279–293, 1994.

85. M. U. Kirschbaum, "The temperature dependence of soil organic matter decomposition, and the effect of global warming on soil organic C storage," Soil Biology and Biochemistry, vol. 27, no. 6, pp. 753–760, 1995.

86. E. G. Jobbágy and R. B. Jackson, "The vertical distribution of soil organic carbon and its relation to climate and vegetation," Ecological Applications, vol. 10, no. 2, pp. 423–436, 2000.

87. S. Long, "Modification of the response of photosynthetic productivity to rising temperature by atmospheric CO_2 concentrations: has its importance been underestimated?" Plant, Cell & Environment, vol. 14, no. 8, pp. 729–739, 1991.

88. T. Ise and P. R. Moorcroft, "The global-scale temperature and moisture dependencies of soil organic carbon decomposition: an analysis using a mechanistic decomposition model," Biogeochemistry, vol. 80, no. 3, pp. 217–231, 2006.

89. D. S. Jenkinson, D. E. Adams, and A. Wild, "Model estimates of CO_2 emissions from soil in response to global warming," Nature, vol. 351, no. 6324, pp. 304–306, 1991.

90. E. D. Schulze and A. Freibauer, "Environmental science: carbon unlocked from soils," Nature, vol. 437, no. 7056, pp. 205–206, 2005.

91. C. Fang, P. Smith, J. B. Moncrieff, and J. U. Smith, "Similar response of labile and resistant soil organic matter pools to changes in temperature," Nature, vol. 433, no. 7021, pp. 57–59, 2005.

92. C. Fang, P. Smith, J. B. Moncrieff, and J. U. Smith, "Corrigendum: Similar response of labile and resistant soil organic matter pools to changes in temperature," Nature, vol. 436, no. 7052, p. 881, 2005.

93. W. Knorr, I. C. Prentice, J. I. House, and E. A. Holland, "Long-term sensitivity of soil carbon turnover to warming," Nature, vol. 433, no. 7023, pp. 298–301, 2005.

94. J. Sanderman, R. G. Amundson, and D. D. Baldocchi, "Application of eddy covariance measurements to the temperature dependence of soil organic matter mean residence time," Global Biogeochemical Cycles, vol. 17, no. 2, pp. 30–1, 2003.
95. P. Dalias, J. M. Anderson, P. Bottner, and M.-M. Coûteaux, "Long-term effects of temperature on carbon mineralisation processes," Soil Biology and Biochemistry, vol. 33, no. 7-8, pp. 1049–1057, 2001.
96. E. A. Davidson and I. A. Janssens, "Temperature sensitivity of soil carbon decomposition and feedbacks to climate change," Nature, vol. 440, no. 7081, pp. 165–173, 2006.
97. M. Albani, D. Medvigy, G. C. Hurtt, and P. R. Moorcroft, "The contributions of land-use change, CO_2 fertilization, and climate variability to the Eastern US carbon sink," Global Change Biology, vol. 12, no. 12, pp. 2370–2390, 2006.
98. X. Xu, W. Liu, and G. Kiely, "Modeling the change in soil organic carbon of grassland in response to climate change: effects of measured versus modelled carbon pools for initializing the Rothamsted carbon model," Agriculture, Ecosystems & Environment, vol. 140, no. 3-4, pp. 372–381, 2011.
99. J. Zhan, H. Yan, B. Chen, J. Luo, and N. Shi, "Decomposition analysis of the mechanism behind the spatial and temporal patterns of changes in carbon bio-sequestration in China," Energies, vol. 5, no. 2, pp. 386–398, 2012.
100. M. Jung, M. Reichstein, P. Ciais et al., "Recent decline in the global land evapotranspiration trend due to limited moisture supply," Nature, vol. 467, no. 7318, pp. 951–954, 2010.
101. H. Bormann, "Sensitivity analysis of 18 different potential evapotranspiration models to observed climatic change at German climate stations," Climatic Change, vol. 104, no. 3-4, pp. 729–753, 2011.
102. J. Lu, G. Sun, S. G. Mcnulty, and D. M. Amatya, "A comparison of six potential evapotranspiration methods for regional use in the southeastern United States," Journal of the American Water Resources Association, vol. 41, no. 3, pp. 621–633, 2005.
103. P. Martin, N. J. Rosenberg, and M. S. McKenney, "Sensitivity of evapotranspiration in a wheat field, a forest, and a grassland to

changes in climate and direct effects of carbon dioxide," Climatic Change, vol. 14, no. 2, pp. 117–151, 1989.

104. M. S. Mckenney and N. J. Rosenberg, Simulating the Impacts of Climate Change on Potential Evapotranspiration: A Comparison of the Climate Sensitivity of Selected Methods, Resource for the Future, Washington, DC, USA, 1991.

105. S. J. Reddy, "Sensitivity of some potential evapotranspiration estimation methods to climate change,"Agricultural and Forest Meteorology, vol. 77, no. 1-2, pp. 121–125, 1995.

106. J. Pastor and W. M. Post, "Response of northern forests to CO_2-induced climate change," Nature, vol. 334, no. 6177, pp. 55–58, 1988.

107. A. M. Solomon, "Transient response of forests to CO_2-induced climate change: simulation modeling experiments in eastern North America," Oecologia, vol. 68, no. 4, pp. 567–579, 1986.

108. J. T. Overpeck, D. Rind, and R. Goldberg, "Climate-induced changes in forest disturbance and vegetation," Nature, vol. 343, no. 6253, pp. 51–53, 1990.

109. P. M. Cox, R. A. Betts, C. D. Jones, S. A. Spall, and I. J. Totterdell, "Acceleration of global warming due to carbon-cycle feedbacks in a coupled climate model," Nature, vol. 408, no. 6809, pp. 184–187, 2000.

110. R. L. Peters and J. D. S. Darling, "The greenhouse effect and nature reserves," BioScience, vol. 35, no. 11, pp. 707–717, 1985.

111. J. M. Melillo, "Vegetation/ecosystem modeling and analysis project: comparing biogeography and biogeochemistry models in a continental-scale study of terrestrial ecosystem responses to climate change and CO_2 doubling," Global Biogeochemical Cycles, vol. 9, no. 4, pp. 407–437, 1995.

112. R. A. Betts, P. M. Cox, S. E. Lee, and F. I. Woodward, "Contrasting physiological and structural vegetation feedbacks in climate change simulations," Nature, vol. 387, no. 6635, pp. 796–799, 1997.

113. R. G. Gallimore and J. E. Kutzbach, "Role of orbitally induced changes in tundra area in the onset of glaciation," Nature, vol. 381, no. 6582, pp. 503–505, 1996.

114. S. Piao, P. Friedlingstein, P. Ciais, L. Zhou, and A. Chen, "Effect of climate and CO_2 changes on the greening of the Northern Hemisphere over the past two decades," Geophysical Research Letters, vol. 33, no. 23, Article ID L23402, 2006.
115. J. E. Lovelock and L. R. Kump, "Failure of climate regulation in a geophysiological model," Nature, vol. 369, no. 6483, pp. 732–734, 1994.
116. R. Costanza, R. D'Arge, R. De Groot et al., "The value of the world's ecosystem services and natural capital," Nature, vol. 387, no. 6630, pp. 253–260, 1997.
117. R. C. Estoque and Y. Murayama, "Examining the potential impact of land use/cover changes on the ecosystem services of Baguio city, the Philippines: a scenario-based analysis," Applied Geography, vol. 35, no. 1-2, pp. 316–326, 2012.
118. Z. F. Bian and Q. Q. Lu, "Ecological effects analysis of land use change in coal mining area based on ecosystem service valuing: a case study in Jiawang," Environmental Earth Sciences, vol. 68, no. 6, pp. 1619–1630, 2013.
119. E. G. Mcpherson, "Environmental benefits and costs of the urban forest," in Proceedings of the 5th National Urban Forest Conference, pp. 52–54, 1992.
120. J. F. Dwyer, E. G. Mcpherson, H. W. Schroeder, and R. A. Rowntree, "Assessing the benefits and costs of the urban forest," Journal of Arboriculture, vol. 18, pp. 227–227, 1992.
121. P. C. Howard, The Economics of Protected Areas in Uganda: Costs, Benefits and Policy Issues, University of Edinburgh, 1995.
122. C. Corvalan, S. Hales, and A. J. McMichael, Mcmichael, Ecosystems and Human Well-Being: Health Synthesis, World Health Organization, 2005.
123. N. A. Chappell, Y. D. Chen, J. Suhaimi, and M. Bonell, "Climate regulation of humid tropical hydrology," in Sustainable Hydrology for the 21st Century. Proceedings of the 10th BHS National Hydrology Symposium, pp. 172–177, British Hydrological Society, Exeter, UK, 2008.
124. Y. D. Chen and N. A. Chappell, "Climate regulation of Southeast Asian hydrology," in Critical States: Environmental Challenges to Development in Monsoon Southeast Asia. Gerakbudaya, Kuala Lumpur, pp. 205–220, 2009.

125. C. Rosenzweig, W. D. Solecki, L. Parshall, M. Chopping, G. Pope, and R. Goldberg, "Characterizing the urban heat island in current and future climates in New Jersey," Environmental Hazards, vol. 6, no. 1, pp. 51–62, 2005.
126. P. Bolund and S. Hunhammar, "Ecosystem services in urban areas," Ecological Economics, vol. 29, no. 2, pp. 293–301, 1999.
127. R. B. Richardson, "Ecosystem services and food security: economic perspectives on environmental sustainability," Sustainability, vol. 2, pp. 3520–3548, 2010.
128. M. V. K. Sivakumar, H. P. Das, and O. Brunini, "Impacts of present and future climate variability and change on agriculture and forestry in the arid and semi-arid tropics," Climatic Change, vol. 70, no. 1-2, pp. 31–72, 2005.
129. Q. Jiang, X. Deng, J. Zhan, and S. He, "Estimation of land production and its response to cultivated land conversion in North China Plain," Chinese Geographical Science, vol. 21, no. 6, pp. 685–694, 2011.
130. Q. L. Shi, Y. Z. Lin, E. P. Zhang, H. M. Yan, and J. Y. Zhan, "Impacts of cultivated land reclamation on the climate and grain production in Northeast China in the future 30 years," Advances in Meteorology, vol. 2013, Article ID 853098, 8 pages, 2013.
131. S. Lautenbach, C. Kugel, A. Lausch, and R. Seppelt, "Analysis of historic changes in regional ecosystem service provisioning using land use data," Ecological Indicators, vol. 11, no. 2, pp. 676–687, 2011.
132. R. Lal, R. F. Follett, B. A. Stewart, and J. M. Kimble, "Soil carbon sequestration to mitigate climate change and advance food security," Soil Science, vol. 172, no. 12, pp. 943–956, 2007.
133. N. Gedney and P. J. Valdes, "The effect of Amazonian deforestation on the northern hemisphere circulation and climate," Geophysical Research Letters, vol. 27, no. 19, pp. 3053–3056, 2000.
134. R. Lal and R. F. Follett, "Soil Carbon Sequestration and the Greenhouse Effect," ASA-CSSA-SSSA, 2009.

Citations

CHAPTER 1

Christian Hof, Miguel B. Arau´jo, Walter Jetz, and Carsten Rahbek, Additive threats from Pathogens, climate and land-use change for global amphibian diversity, doi:10.1038/nature10650.

CHAPTER 2

R. B. Singh and Chenchen Shi, "Advances in Observation and Estimation of Land Use Impacts on Climate Changes: Improved Data, Upgraded Models, and Case Studies," Advances in Meteorology, vol. 2014, Article ID 748169, 7 pages, 2014. doi:10.1155/2014/748169.

CHAPTER 3

Yanhu He, Kairong Lin, and Xiaohong Chen, "Effect of Land Use and Climate Change on Runoff in the Dongjiang Basin of South China," Mathematical Problems in Engineering, vol. 2013, Article ID 471429, 14 pages, 2013. doi:10.1155/2013/471429.

CHAPTER 4

Feng Wu, Jinyan Zhan, Hongbo Su, Haiming Yan, and Enjun Ma, "Scenario-Based Impact Assessment of Land Use/Cover and Climate Changes on Watershed Hydrology in Heihe River Basin of Northwest China,"Advances in Meteorology, Article ID 410198, in press.

CHAPTER 5

Xiangzheng Deng, Chunhong Zhao, and Haiming Yan, "Systematic Modeling of Impacts of Land Use and Land Cover Changes on Regional Climate: A Review," Advances in Meteorology, vol. 2013, Article ID 317678, 11 pages, 2013. doi:10.1155/2013/317678.

CHAPTER 6

Zhihui Li, Xiangzheng Deng, Fang Yin, and Cuiyuan Yang, "Analysis of Climate and Land Use Changes Impacts on Land Degradation in the North China Plain," Advances in Meteorology, Article ID 976370, in press.

CHAPTER 7

Jiming Jin, Shihua Lu, Suosuo Li, and Norman L. Miller, "Impact of Land Use Change on the Local Climate over the Tibetan Plateau," Advances in Meteorology, vol. 2010, Article ID 837480, 6 pages, 2010. doi:10.1155/2010/837480.

CHAPTER 8

Tao Zhang, Jinyan Zhan, Juan Huang, Rui Yu, and Chenchen Shi, "An Agent-Based Reasoning of Impacts of Regional Climate Changes on Land Use Changes in the Three-River Headwaters Region of China," Advances in Meteorology, vol. 2013, Article ID 248194, 9 pages, 2013. doi:10.1155/2013/248194.

CHAPTER 9

Zhihui Li, Xiangzheng Deng, Jikun Huang, Rongrong Zhang, and Juan Huang, "Critical Studies on Integrating Land-Use Induced Effects on Climate Regulation Services into Impact Assessment for Human Well-Being," Advances in Meteorology, vol. 2013, Article ID 831250, 14 pages, 2013. doi:10.1155/2013/831250.

Index